Advanced AutoCAD 2010

EXERCISE WORKBOOK

by
Cheryl R. Shrock

Professor
Drafting Technology
Orange Coast College, Costa Mesa, Ca.
Autodesk Authorized Author

INDUSTRIAL PRESS
New York

Industrial Press Inc.
989 Avenue of the Americas
New York, NY 10018

10 9 8 7 6 5 4 3 2 1

I retired from the classroom this year.

I will miss the interaction with my students. So I am pleased to be able to continue assisting students, in the education of AutoCAD, through the writing of my workbooks. I will continue as long as the workbooks are helpful.

*So let me know….**learnacad@yahoo.com***

AutoCAD Books by Cheryl R. Shrock:

Beginning AutoCAD **2006**   ISBN 978-08311-3213-2
Advanced AutoCAD **2006**   ISBN 978-08311-3214-9

Beginning AutoCAD **2007**   ISBN 978-08311-3302-3
Advanced AutoCAD **2007**   ISBN 978-08311-3303-0

Beginning AutoCAD **2008**   ISBN 978-0-8311-3341-2
Advanced AutoCAD **2008**   ISBN 978-0-8311-3342-9

Beginning AutoCAD **2009**   ISBN 978-0-8311-3359-7
Advanced AutoCAD **2009**   ISBN 978-0-8311-3360-3

Beginning AutoCAD **2010**   ISBN 978-0-8311-3404-4
Advanced AutoCAD **2010**   ISBN 978-0-8311-3400-6

AutoCAD Pocket Reference   ISBN 978-0-8311-3263-7
2006, Releases 2006-2004

AutoCAD Pocket Reference   ISBN 978-0-8311-3328-3
2007 Releases 2007-2004

AutoCAD Pocket Reference   ISBN 978-0-8311-3354-2
2008 Releases 2008-2006

AutoCAD Pocket Reference   ISBN 978-0-8311-3384-9
4th Edition, Release 2009 or later

For information about these books visit www.industrialpress.com

For information about Cheryl Shrock's online courses,
visit www.shrockpublishing.com

Table of Contents

Lesson 12

Lesson 13

Lesson 14

Lesson 15

Lesson 16

Lesson 17

INTRODUCTION

About this workbook

This workbook is designed to <u>follow</u> the ***Beginning AutoCAD 2010, Exercise Workbook***. It is excellent for classroom instruction or self-study. There are 20 lessons and 3 *on-the-job* type projects in Architectural, Electro-mechanical and Mechanical.

12 Lessons continue your education in basic 2D commands.

8 Lessons introduce you to many basic 3D commands.

Each lesson starts with step-by-step instructions followed by exercises designed for practicing the commands you learned within that lesson. The *on-the-job* projects are designed to give you more practice in your desired field of drafting.

> *Important*
> Files ***2010-Workbook Helper*** and ***2010-3D Demo*** *should be downloaded from our website:* **www.shrockpublishing.com**

AutoCAD 2010 vs. AutoCAD LT 2010

The LT version of AutoCAD has approximately 80 percent of the capabilities of the full version. It was originally created to be installed on the small hard drives that Laptops used to have. Hence, the name LT. In order to reduce the size of the program AutoCAD removed some of the high-end capabilities, such as Solid Modeling. Through out the workbook I point out which commands are not available to LT users. Consider this an opportunity to see the commands that you are missing and you can determine if you feel it necessary to upgrade.

About the Author

Cheryl R. Shrock is a retired Professor and Chairperson of Computer Aided Design at Orange Coast College in Costa Mesa, California. She is also an Autodesk® registered author. Cheryl began teaching CAD in 1990. Previous to teaching, she owned and operated a commercial product and machine design business where designs were created and documented using CAD. This workbook is a combination of her teaching skills and her industry experience.

The author can be contacted for questions or comments at: *learnacad@yahoo.com*

 "Sharing my industry and CAD knowledge has been the most rewarding experience of my career. Students come to learn CAD in order to find employment or to upgrade their skills. Seeing them actually achieve their goals, and knowing I helped, is a real pleasure. If you read the lessons and do the exercises, I promise, you will not fail."

Cheryl R. Shrock

CONFIGURING YOUR SYSTEM

Note: If you have already configured your system for the 2010 "Beginning" Workbook you may skip to Lesson 1.

While you are using this workbook it is necessary for you to make some simple changes to your configuration so our configurations are the same. This will ensure that the commands and exercises work as expected. The following instructions will guide you through those changes.

1. Start AutoCAD®

2. Type: **_options_** then press the **<enter>** key. (not case sensitive)

   Depending how your software is set up, the text that you type will either appear on the <u>Command Line,</u> as shown below on the left, or in the <u>Dynamic Input box</u>, as shown below on the right. Either is fine for now and will be discussed later.

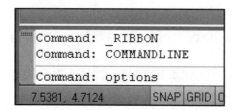

Command Line

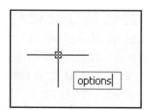

Dynamic Input

NOTE:

AUTOCAD LT USERS:
You may find that some of the settings appear slightly different.
But they are mostly the same.

Configuration Settings

3. Select the *Display* tab and change the settings on your screen to match the dialog box below.

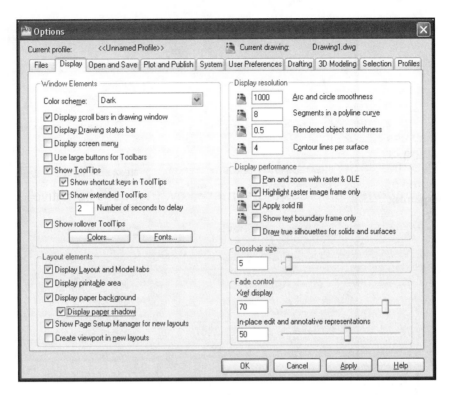

4. Select the *Open and Save* tab and change the settings on your screen to match the dialog box below.

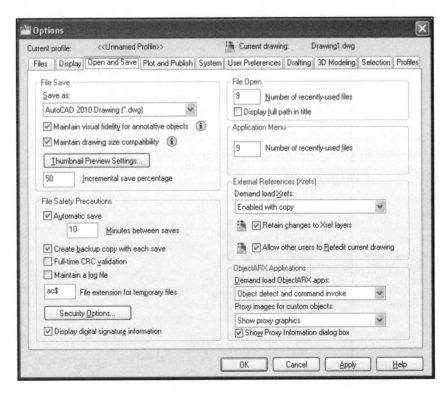

Configuration Settings

5. Select the *Plot and Publish* tab and change the settings on your screen to match the dialog box below.

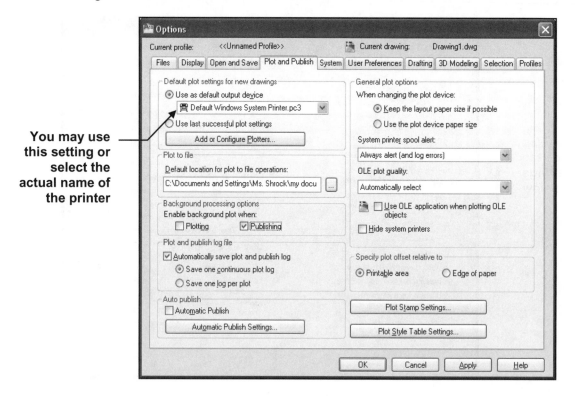

You may use
this setting or
select the
actual name of
the printer

6. Select the *System* tab and change the settings on your screen to match the dialog box below.

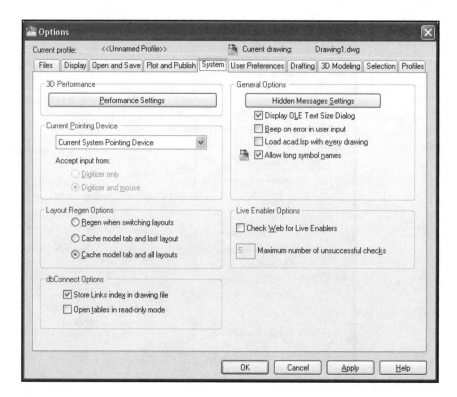

Configuration Settings

7. Select the *User Preferences* tab and change the settings on your screen to match the dialog box below.

Rt-click Cust.
Select this button and change the settings shown below

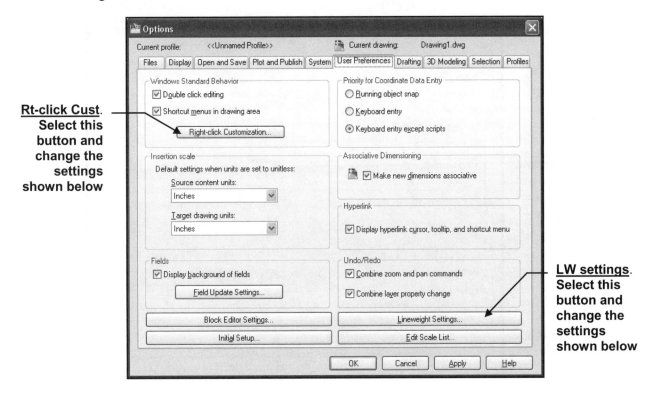

LW settings.
Select this button and change the settings shown below

8. After making the setting changes shown below select **Apply & Close** button.

Right-click Customization

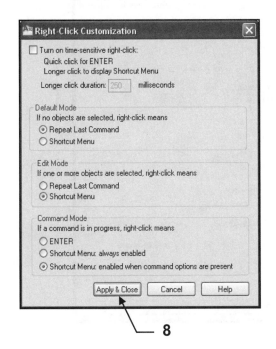

8

Lineweight Settings

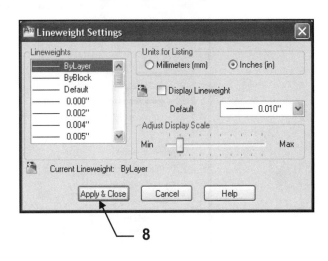

8

Configuration Settings

9. Select the **Drafting** tab and change the settings on your screen to match the dialog box below.

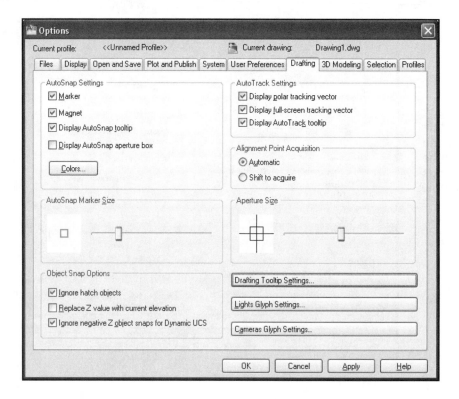

10. Select the **Selection** tab and change the settings on your screen to match the dialog box below. (Note: 3D Modeling tab was skipped)

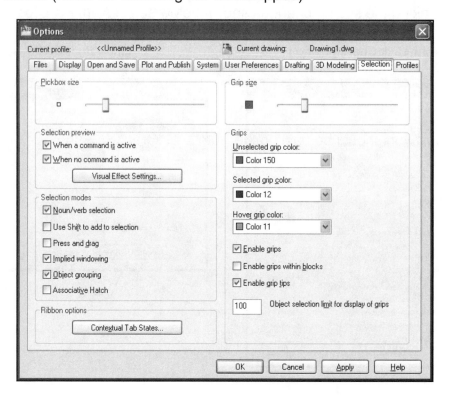

Configuration Settings

11. Select the *Apply* button.

12. Select the *OK* button.

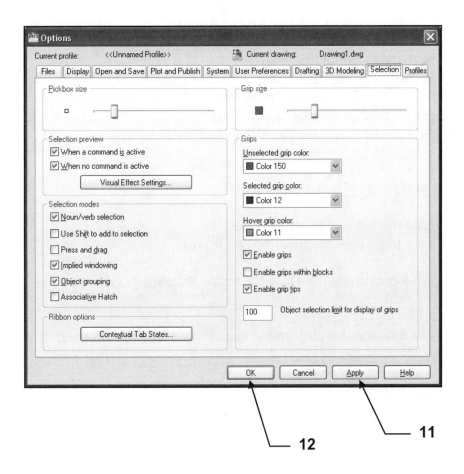

13. Now you should be back to the AutoCAD screen.

Customizing your Wheel Mouse

A Wheel mouse has two or more buttons and a small wheel between the two topside buttons. The default functions for the two top buttons and the Wheel are as follows:
Left Hand button is for **input** and can't be reprogrammed.
Right Hand button is for **Enter** or the **shortcut menu**.
The Wheel may be used to Zoom and Pan or Zoom and display the Object Snap menu. You will learn more about this later.

The following describes how to select the Wheel functions. After you understand the functions, you may choose to change the setting.
To change the setting you must use the **MBUTTONPAN** variable.

MBUTTONPAN setting 1: (Factory setting)

ZOOM   Rotate the wheel forward to zoom in
                      Rotate the wheel backward to zoom out

ZOOM   Double click the wheel to view entire drawing
EXTENTS

PAN   Press the wheel and drag the mouse to move the drawing on the screen.

MBUTTONPAN setting 0:

ZOOM   Rotate the wheel forward to zoom in
                      Rotate the wheel backward to zoom out

OBJECT   Object Snap menu will appear when you press the wheel
SNAP

To change the setting:

1. Type: **mbuttonpan <enter>**
2. Enter **0** or **1 <enter>**

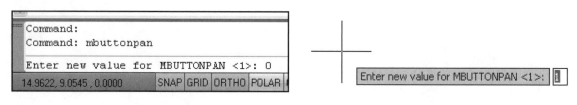

Command Line   **Dynamic Input**

AutoCAD 2010 System Requirements

AutoCAD 2D and AutoCAD LT – <u>32-bit</u>

Operating system:
    Microsoft Windows Vista, any of the versions
    Microsoft Windows XP Home or Professional SP 2

Browser
    Microsoft Internet Explorer 6.0 or later

Memory and Hard Disk Space
    XP-1 GB of RAM minimum
    Vista - 2GB of Ram minimum
    750MB free hard disk space for installation

Hardware (required)
    XP - AMD 64-based or Intel Pentium 4, 1.6 Ghz or higher
    Vista - AMD 64-based or Intel Pentium 4, 3.0 Ghz or higher
    Pointing Device - MS-Mouse compatible
    Display - 1024 X 768 VGA (True color)
    DVD/CD-ROM drive for initial installation only

> **Note:**
> Whether the Windows operating system is the 32-bit or the 64-bit version is automatically detected when installing AutoCAD. The appropriate version of AutoCAD will be installed. The 32-bit version of AutoCAD cannot be installed on a 64-bit version of Windows.

AutoCAD <u>2D</u> – <u>64-bit additional</u> requirements

Operating system:
    Microsoft Windows Vista, 64-bit
    Microsoft Windows XP Professional x64

Memory and Hard Disk Space
    XP Pro - 1GB RAM
    Vista 64-bit – 2GB RAM

Hardware (required)
    Intel 64-based or AMD 64 Processor

AutoCAD <u>3D</u> – 32 or 64-bit <u>additional minimum</u> requirements

Operating system:
    Microsoft Windows Vista
    Microsoft Windows XP Professional

Memory and Hard disk space
    2GB RAM
    2GB free hard disk
    1GB free hard disk for installation (in addition to the 2GB free hard disk above)

Hardware (required)
    Intel Pentium 4.0 GHz Processor or AMD Athion processor, 2.2 Ghz or greater or
    Intel or AMD Dual Core processor, 1.6 Ghz or greater.
    XP Pro - 1280 X 1024 32-bit color video display adapter, True Color, 128 MB, or Direct3D
        capable workstation class graphics card.
    Vista - Direct3D capable workstation class graphics card with 128MB
        1024 X 768 VGA with True Color (minimum)

Go to Autodesk.com to verify graphics hardware for use with AutoCAD 2010

Notes:

LEARNING OBJECTIVES

After completing this lesson you will be able to:

1. Control the display of Multiple Drawings
2. Use the Single-drawing Compatibility mode
3. Test your current AutoCAD skills
4. Plot from model space.

Note:
This lesson should be used to determine whether or not you are ready for this level of instruction. If you have difficulty creating Exercises 1A, 1B and 1C you should consider reviewing the "Exercise Workbook for Beginning AutoCAD 2010" before going on to Lesson 2.

LESSON 1

DISPLAY MULTIPLE DRAWINGS

AutoCAD has an option that allows you to control the number of drawings open at the same time. This option is **"Single-drawing compatibility mode".** If you have "Single-drawing compatibility mode" **ON**, only <u>one</u> drawing can be open on the screen. If you have "Single-drawing compatibility mode" **OFF**, <u>multiple</u> drawings may be open on the screen.

In AutoCAD 2010 this options default setting is OFF. This means that you can have multiple drawings open on the screen at the same time. In the Beginning workbook I instruct you to set this option to ON because I thought it would be less confusing if you had only one drawing open on the screen at anytime. But now you are in the Advanced workbook and you may decide how you want this option set.

HOW TO CONFIGURE SINGLE-DRAWING COMPATIBILITY MODE.

1. Type: **SDI <enter>**
2. Type: **0 <enter>**
   (1 = only one drawing displayed   0 = Multiple drawings displayed)

HOW TO VIEW <u>EACH</u> OPEN DRAWING.

Option 1. Select the **View tab / Windows panel** and select **Switch Windows ▼**

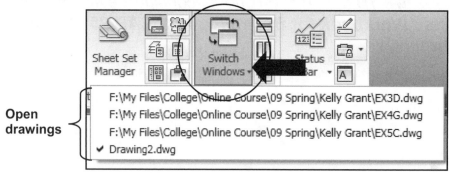

Open drawings

Option 2. Select the **Application Menu** and **Open Documents button**

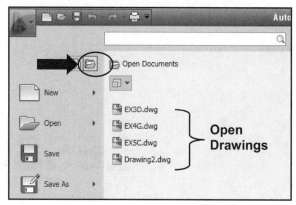

Option 3.
Select **Ctrl + Tab** keys to toggle between drawings.

DISPLAY MULTIPLE DRAWINGS....continued

HOW TO VIEW <u>ALL</u> OPEN DRAWINGS.
You may display all of the open drawings as **cascading**, **vertically** or **horizontally** tiled.

1. Select the **View tab / Window panel**

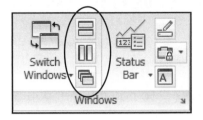

Horizontal

Drawings are arranged in horizontal non-overlapping tiles

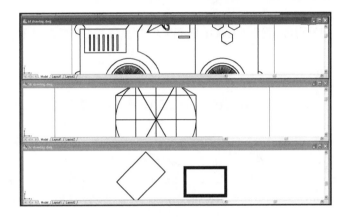

Vertical

Drawings are arranged in vertical non-overlapping tiles

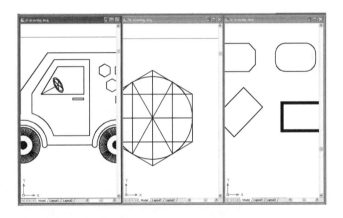

Cascade

Drawings are overlapped but the title bars are visible.

WARM UP DRAWINGS

The following drawings have been included for two purposes:

First purpose: To make sure you remember the commands taught in the Beginning Workbook and to get you prepared for the new commands in this Advanced workbook.

Second purpose: To confirm that you are ready for this level of instruction.

IMPORTANT, PLEASE READ THE FOLLOWING:

*This workbook assumes you already have enough basic AutoCAD knowledge to easily complete Exercises 1A, 1B and 1C. If you have difficulty with these exercises, you should consider reviewing "**Exercise Workbook for Beginning AutoCAD**".   If you try to continue without this knowledge you will probably get confused and frustrated. It is better to have a good solid understanding of the basics before going on to Lesson 2.*

PRINTING

Exercises 1A, 1B and 1C should be printed from Model space on any letter size printer. Instructions are on page 1- 8.

EXERCISE 1A

INSTRUCTIONS:

1. Start a NEW file using **2010-Workbook Helper.dwt** (Refer to Intro-1)
2. Draw the objects below using layers Object Line and Centerline.
3. Do not dimension
4. Save as **EX1A**
5. Plot using the instructions on page 1-8

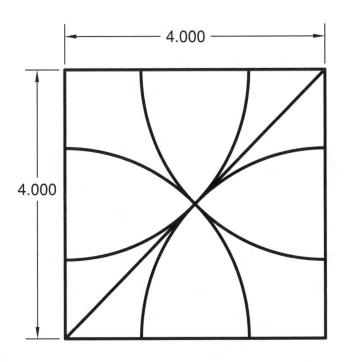

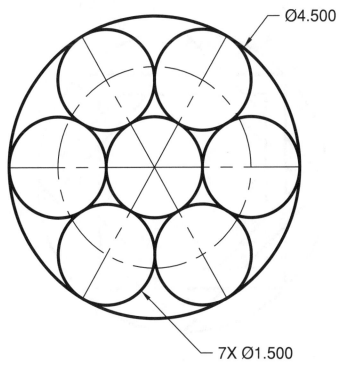

Ø4.500

7X Ø1.500

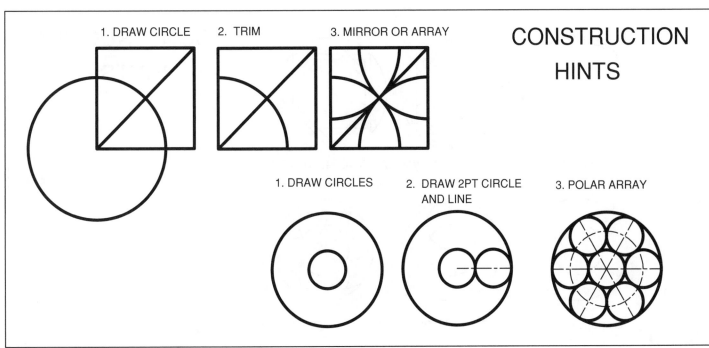

1. DRAW CIRCLE   2. TRIM   3. MIRROR OR ARRAY

CONSTRUCTION
HINTS

1. DRAW CIRCLES   2. DRAW 2PT CIRCLE AND LINE   3. POLAR ARRAY

EXERCISE 1B

INSTRUCTIONS:
1. Start a NEW file using **2010-Workbook Helper.dwt** (Refer to Intro-1)
2. Draw the objects below using layers Object Line
3. Do not dimension
4. Save as **EX1B**
5. Plot using the instructions on page 1-8

Ø4.500 —   OFFSET .300

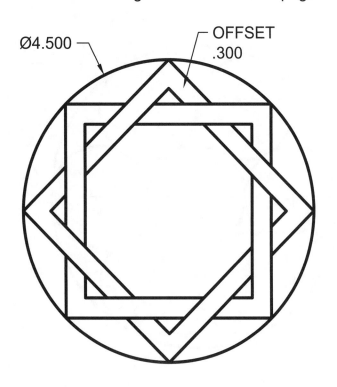

Ø4.500 —   OFFSET.300

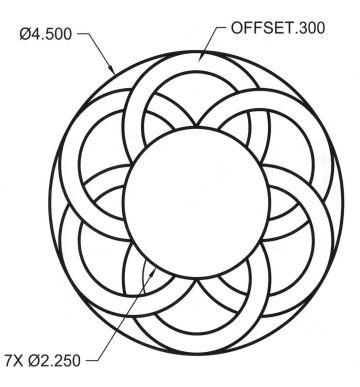

7X Ø2.250

CONSTRUCTION HINTS

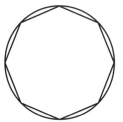

1. 8 Sided Polygon

2. 4 Sided Polygons

3. Offset and Trim

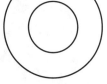

1. Circle, Cen / Rad

2. Circle, 2P

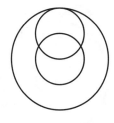

3. Array, Offset, Trim

EXERCISE 1C

INSTRUCTIONS:

1. Start a **NEW** file using **2010-Workbook Helper.dwt** (Refer to Intro-1)
2. Draw the objects below using layers Object Line, Centerline and Dimension
3. Dimension using "Class Style".
4. Save as **EX1C**
5. Plot using the instructions on page 1-8

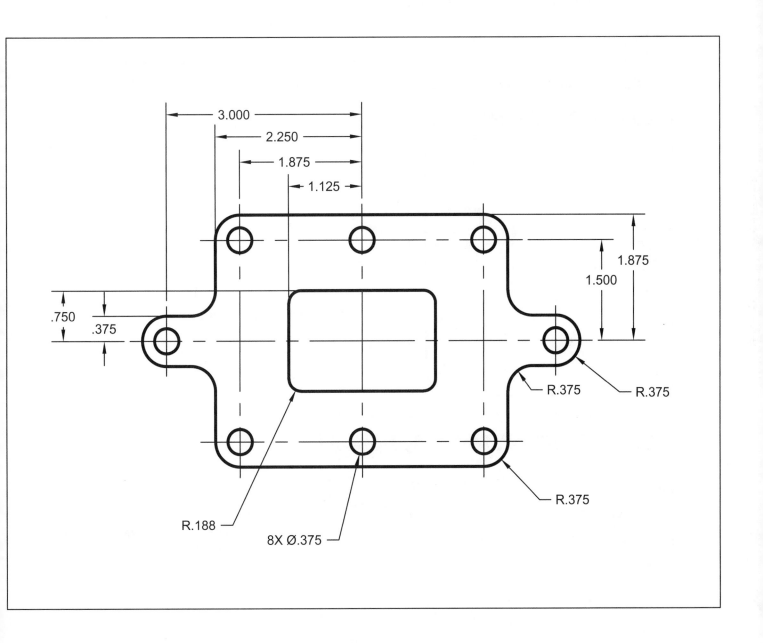

Basic Plotting from Model Space

1. **Important:** Open the drawing you want to plot.
2. Select: **Zoom / All** to center the drawing within the plot area.
3. Select the Plot command using one of the following;

   Quick Access toolbar =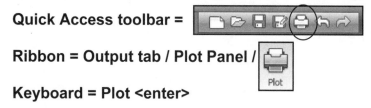

   Ribbon = Output tab / Plot Panel / [Plot]

   Keyboard = Plot <enter>

The Plot –Model dialog box will appear.

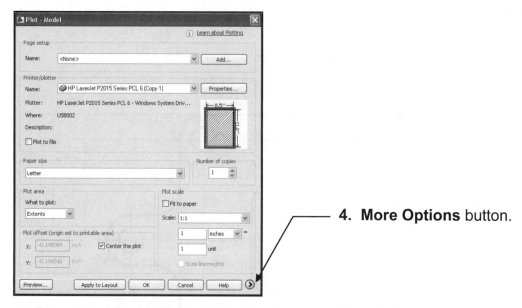

4. More Options button.

4. Select the **"More Options"** button to expand the dialog box.

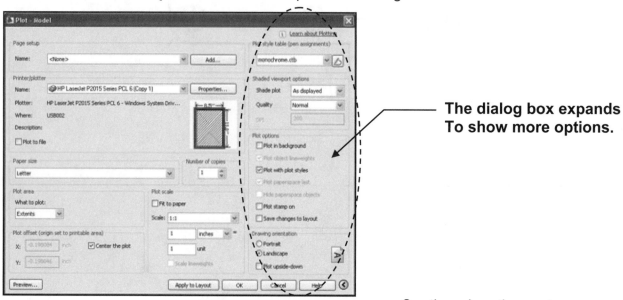

The dialog box expands
To show more options.

Continued on the next page...

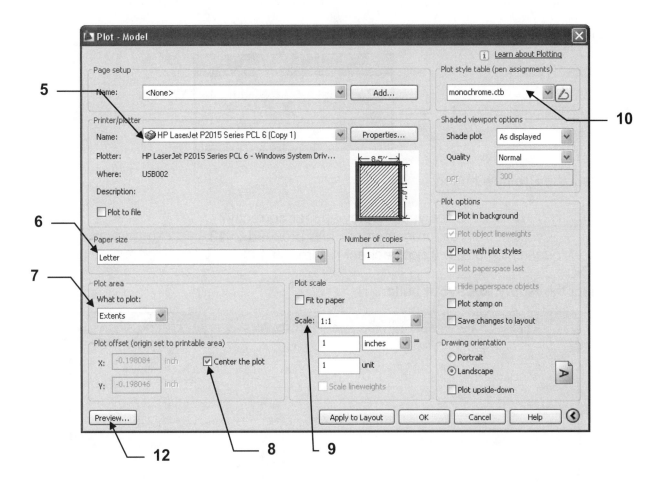

5. Select your printer from the drop down list or "Default windows system printer.pc3"
   Note: If your printer is not shown in the list you should configure your printer within
   AutoCAD. This is not difficult. Refer to Appendix-A for step by step instructions.

6. Select the Paper size :**Letter**.

7. Select the Plot Area **Extents**

8. Select the Plot Offset **Center the Plot**

9. Select Plot Scale **1 : 1**

10. Select the Plot Style table **Monochrome.ctb** (for all black)
    Acad.ctb (for color)

Continued on the next page...

11. If the following box appears, select **Yes**

11 ——

12. Select the **Preview** button.

Does your display appear correctly?
If yes, press <enter> and proceed to 13.
If not, recheck 1 through 11.

13. Select the **Apply to Layout** button.

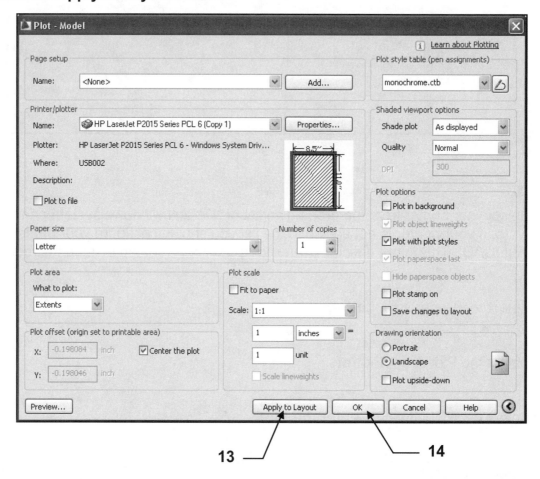

13 —— 14

14. Select the **OK** button to send the drawing to the printer or <u>select Cancel if you do not want to print the drawing at this time.</u>

15. Save the entire drawing again
The plot setting selections will be saved to the drawing.

LEARNING OBJECTIVES

After completing this lesson, you will be able to:

1. Customizing the Workspace
2. Customize the Ribbon
3. Customize the Quick Access Toolbar
4. Select, Delete, Export and Import a Workspace
5. Customize the Status Bar

LESSON 2

Customizing the Workspace

At this stage of your AutoCAD education you are probably not too interested in customizing AutoCAD. But I would like to at least introduce you to the **Custom User Interface** to familiarize you with a few of the customizing options that are available. It is relatively easy to use and sometimes helpful. If you find it intriguing you may explore further using the AutoCAD "Help" system discussed in Lesson 1 in the Beginning workbook.

Note: Those of you that are familiar with the customizing process of previous AutoCAD versions should note that the <u>Custom User Interface</u> file structure replaces the <u>MNU/MNS</u> file structure. If you are someone that enjoys customizing AutoCAD, you will find this tool will make customizing easier because all customizing is done within AutoCAD. You do not have to venture outside of AutoCAD.

Start by opening the **Customize User Interface** dialog box using one of the following:

RIBBON = MANAGE TAB / CUSTOMIZATION PANEL /

KEYBOARD = CUI

The following dialog box should appear.

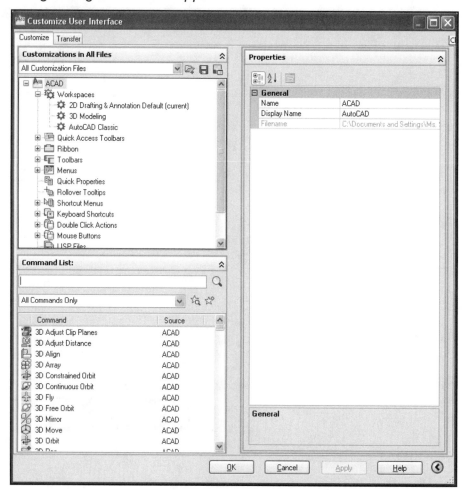

Continued on the next page...

Creating a New Workspace

Have you found yourself wishing that you could move some of the commands from one Ribbon Panel to another? Or possibly Add, Remove or Move the Tabs. Well you can. You can create a **Workspace** all of your own very easily.

Important: Before you begin customizing your workspace it is important to create a Duplicate Workspace rather than modify an existing workspace. This will allow you to revert back to the original AutoCAD default workspace appearance.

Don't be afraid to experiment. I will also show you how to easily return to the AutoCAD default workspace.

1. Open the **Customize User Interface** dialog box as shown on the previous page.

2. Duplicate the **2D Drafting & Annotation** workspace as follows:
   A. Right click on **2D Drafting & Annotation** and select **Duplicate.**

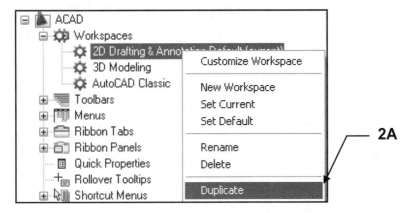

3. Rename the duplicate workspace as follows:
   A. Right click on **Copy of 2D Drafting & Annotation 1** and select **Rename**.

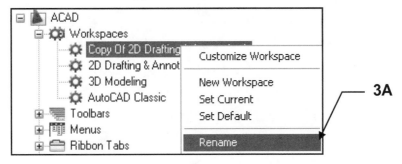

4. Enter the New Workspace name **Class Workspace Demo** <enter>

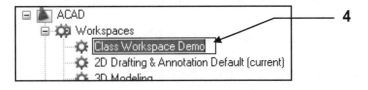

Continued on the next page...

Creating a New Workspace....continued

5. Right Click on the New Workspace **Class Workspace Demo** and select **Set current.**

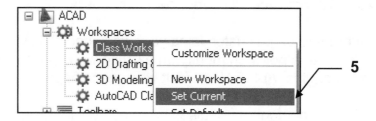

5

6. The New Workspace should now be displayed as shown below.

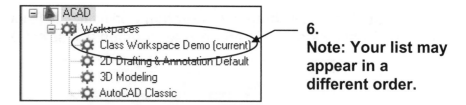

6.
Note: Your list may
appear in a
different order.

7. Select **Apply** and **OK** buttons located at the bottom of the CUI dialog box.

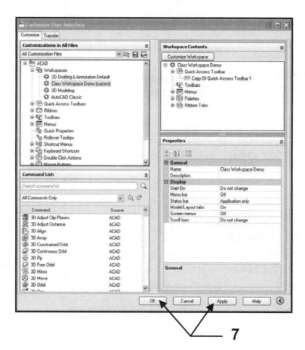

7

Now you have a workspace to customize. The following pages will guide you through:

1. **Creating a new Ribbon tab**

2. **Adding a new Ribbon Panel to the new tab.**

3. **Adding Commands to the new Panel.**

Create a Ribbon tab

1. Open the Customize User Interface dialog box (Refer to page 2-2)

2. Select **Class Work Space Demo**

3. Click on the **[+]** beside Ribbon to expand.

4. Right Click on **Tabs**

5. Select **New Tab**

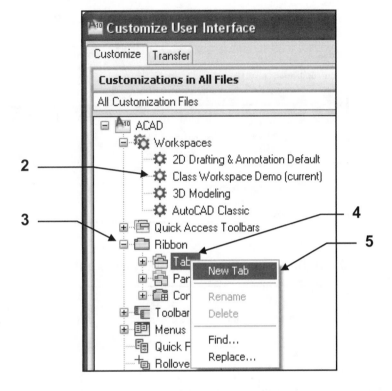

6. Enter a name for the new tab and <enter>. (Note: I entered Class tab)

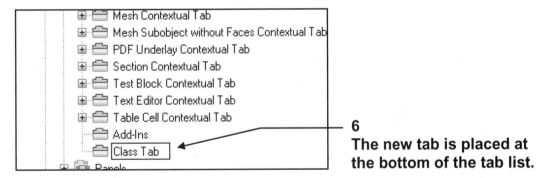

6
The new tab is placed at the bottom of the tab list.

7. Select the **Apply** and then **OK** button at the bottom of the Customize User Interface box.

Note:
New Tabs are not automatically added to the workspace.
You must tell AutoCAD that you want to display this new tab in your workspace.
Follow the steps on the following page to "Add a Tab to a Workspace".

2-5

Add a Ribbon Tab to a Workspace

1. Open the Customize User Interface dialog box (Refer to page 2-2)

2. Select **Class Work Space Demo**

3. Click on **Customize Workspace** button. (It will change to **Done**)

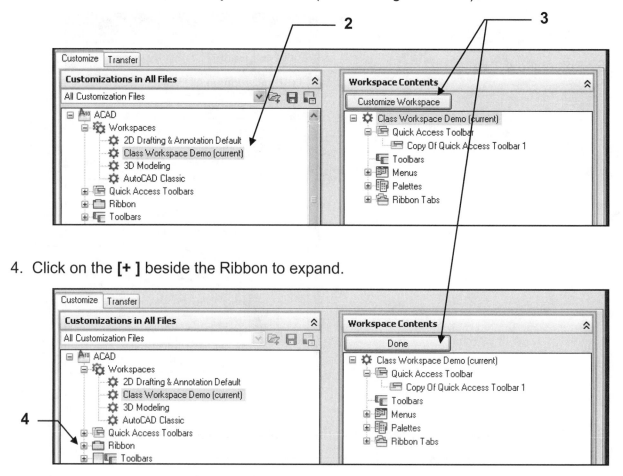

4. Click on the **[+]** beside the Ribbon to expand.

5. Click the check box beside the ribbon tab.

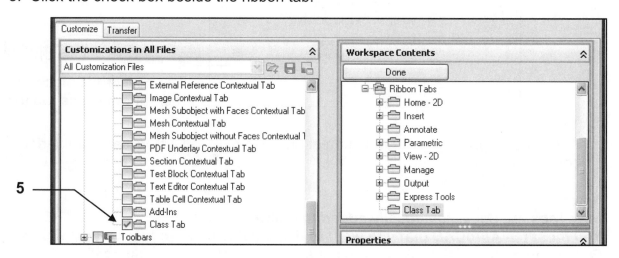

Continued on the next page...

Add a Ribbon Tab to a Workspace....continued

Notice the **Class tab** was added to the **Workspace Contents** also.

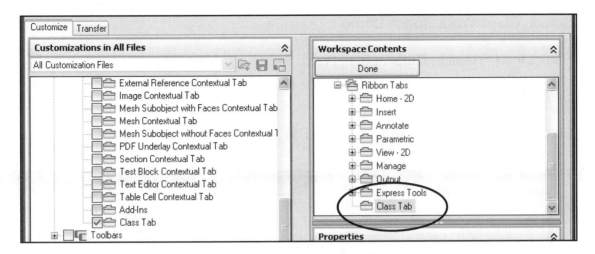

6. Click on the **Done** button.

7. Select the **Apply** and **OK** button.

The New Tab should now appear in the Ribbon.

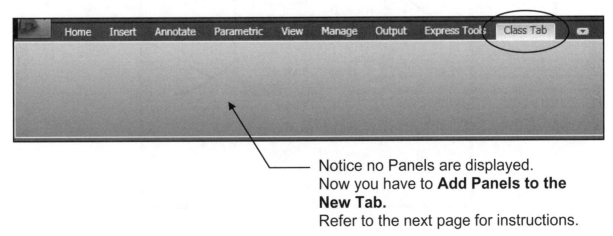

Notice no Panels are displayed.
Now you have to **Add Panels to the New Tab.**
Refer to the next page for instructions.

Add a Ribbon Panel to a tab

1. Open the Customize User Interface dialog box. (Refer to page 2-2)

2. Select **Class Work Space Demo**

3. Select the **[+]** beside **Ribbon** to expand.

4. Select the **[+]** beside **Tabs** to expand.

5. Select the **[+]** beside **Panels** to expand.

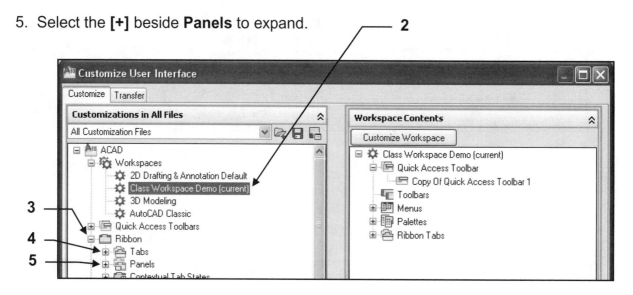

6. Scroll down the list of Panels and select the one you wish to add to the tab.

7. **Right click** on the **Panel** and select **Copy**. from the menu.

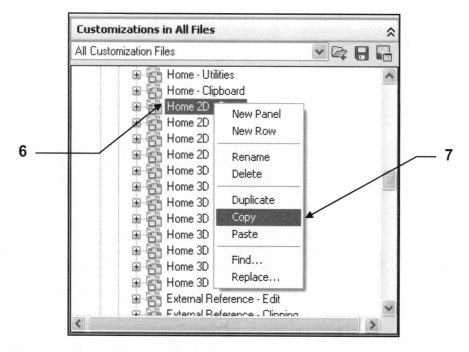

Continued on the next page...

Add a Ribbon Panel to a tab....continued

8. Find the Tab to which you wish to add a panel.

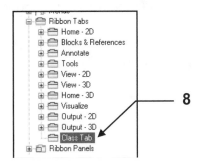

8

9. **Right Click** on the **Tab name** and select **Paste.**

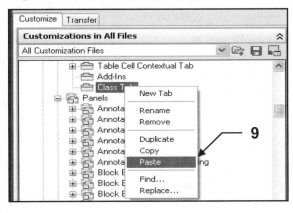

9

10. The Panel should now be listed under **Class Tab**

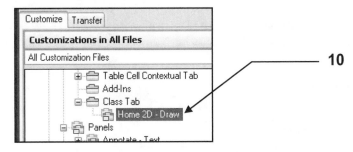

10

11. Select **Apply** and **OK** buttons located at the bottom of the Customize User Interface box.

12. Select the **Tab**. (The added Panel should appear)

12

Create a New Ribbon Panel

The previous page showed you how to add an **existing** Ribbon Panel to a tab. The following will guide you through creating a New Ribbon Panel to which you will add commands and then you will add it to a Ribbon tab.

1. Open the Customize User Interface dialog box (Refer to page 2-2)

2. Select **Class Work Space Demo**

3. Select the **[+]** beside **Ribbon** to expand.

4. Right Click on **Panels**

5. Select **New Panel**

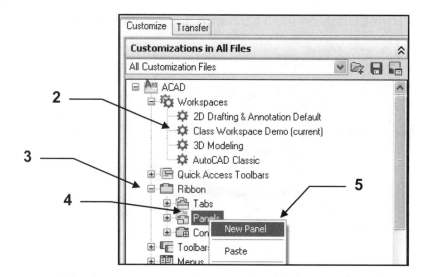

6. Enter a name for the new panel and <enter>. (Note: I entered **My Panel**)

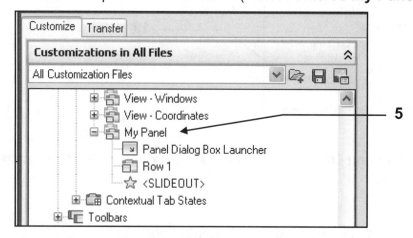

7. Select **Apply** and **OK** buttons.

You will now add commands to the new panel.
*Refer to the next page for **Adding commands to a panel**.*
After adding commands to the panel you will add the new panel to a tab.

Add a Command to a Ribbon Panel

1. Open the Customize User Interface dialog box (Refer to page 2-2)

2. Select **Class Work Space Demo**

3. Select the **[+]** beside **Ribbon** to expand.

4. Select the **[+]** beside **Panels** to expand.

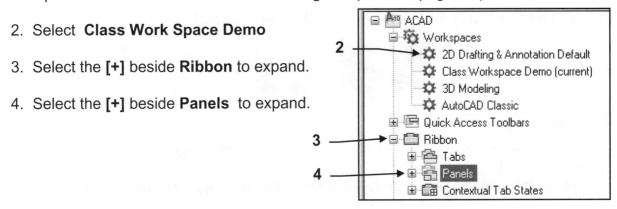

5. Select the **[+]** beside the **Panel name** to which you want to **add commands**.

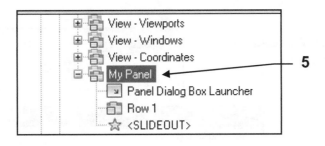

5. Locate the command, that you wish to add, in the Command List area.

6. Click, drag and drop the command on Row 1.

7. Now refer back to page 2-8 for instructions to **Add a Panel to a Tab**

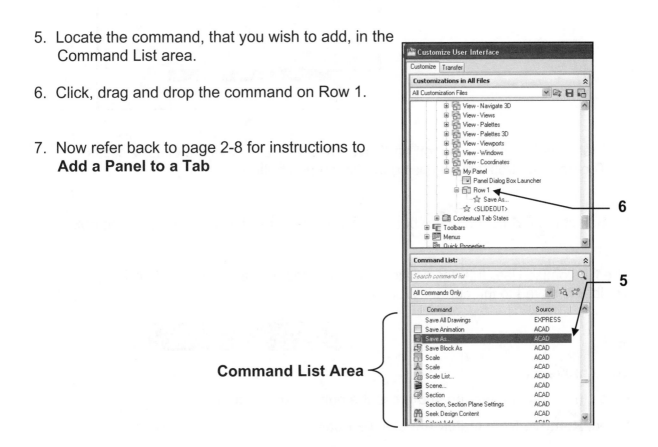

Customize the Quick Access Toolbar

Customize the Quick Access Toolbar

You can add tools with the Customize User Interface dialog box.

For Example:

I like students new to AutoCAD to use the "**Save As**" commandl rather than the "**Save**" command. So I add the "**Save As**" tool to the Quick Access Toolbar. If you would like to add the "**Save As**" tool to your Quick Access Toolbar follow the steps below.

1. Place the Cursor on the Quick Access Toolbar and press the right mouse button.

2. Select **"Customize Quick Access Toolbar..."** from the pop up menu.

3. Scroll through the list of Commands to "**Save As**..."

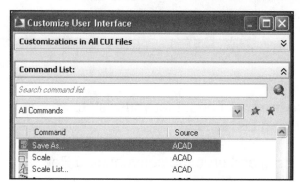

4. Press the Left mouse button on "Save As..." and drag it to the Quick Access Toolbar and drop it by releasing the left mouse button.

5. Select the **OK** button at the bottom of the Customize User Interface dialog box.

The Customize User Interface dialog box will disappear and the new Quick Access Toolbar is saved to the current Workspace.

To Remove a tool:

Place the cursor on the top and press the right mouse button.

Select **Remove from Quick Access Toolbar.**

How to select an existing Workspace

1. Select the **Workspace Switching** down arrow located in the lower right corner.

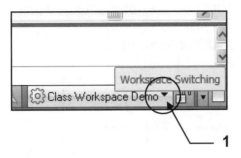

2. Select the Workspace that you prefer.

   The **Current** Workspace is designated with a checked box.

Select the **2D Drafting & Annotation** workspace if you choose to return to the AutoCAD default workspace used in this workbook.

How to Delete a Workspace

1. Select the **Customize User Interface**.

2. **Right click** on **any** Workspace that you **do not** want to delete.
   Select **Set Current.**
   (You can not delete a workspace if it is "current".)

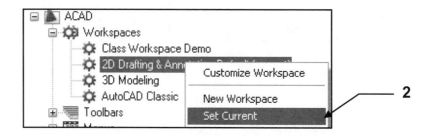

3. **Right click** on the Workspace that you wish to Delete and select **Delete**.

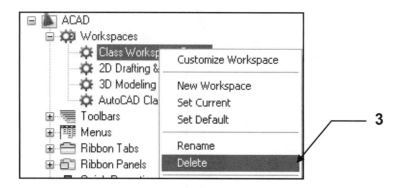

4. Select **Yes** if you really want to delete the workspace.

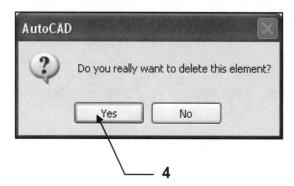

Export a Workspace

Now that you have created your workspace you may wish to **Export** the workspace so you may **Import** the workspace onto another computer.

1. Type **CUIEXPORT <enter>**

2. Select the Workspace you wish to export.

Notice the Transfer tab has been selected

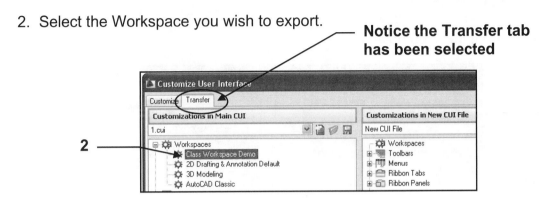

3. Select **Save As** from the drop down list.

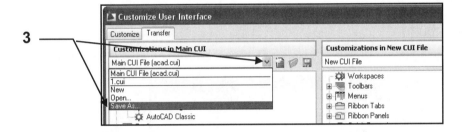

4. Locate where you wish to save the workspace file.
   (Note: This is a fairly large file so make sure you have data area available.)

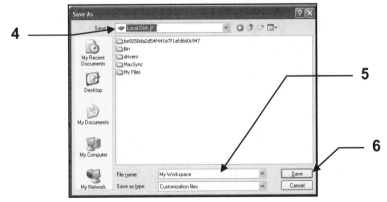

5. Enter a name

6. Select the **Save** button.

Your customized workspace is now saved and can be imported onto another computer. Import instructions are on the next page.

Import a Workspace

Now that you have created your workspace you may wish to **Import** the workspace onto another computer. Note: you should be importing the workspace into a computer that does not have the workspace that you wish to import.

1. Type **CUIIMPORT <enter>**

2. Select **Open** from the drop down list.

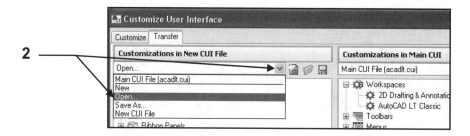

3. Locate the previously saved workspace file.

4. Select the **.cuix** file.

5. Select the **Open** button.

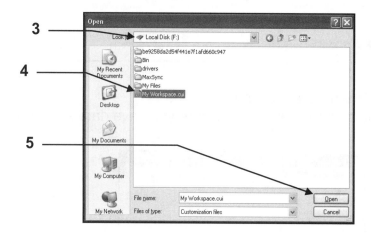

6. **Set Current** if you wish to use it. (Refer to page 2-4)

7. Select the **Apply** and **OK** button located at the bottom of the dialog box.

How to customize the Status Bar

AutoCAD allows you to decide what you would like displayed on the **Status Bar**. Customizing the Status Bar is strictly personal preference. You decide.

1. Right click on the **Status Bar**.

A pops up menu appears listing all tool options.

A check mark means the tool is displayed. If you click on any of the tool names it will uncheck and will not be displayed. Repeat that   selection to bring it back.

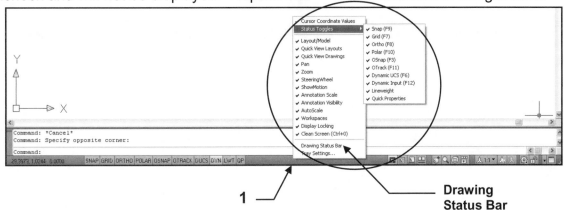

1

Drawing Status Bar

Drawing Status Bar

I like to display the Drawing Status Bar. It basically separates some of the drawing tools from the Status Bar and relocates them directly under the drawing area.

**Drawing Status Bar
ON**

**Drawing Status Bar
OFF**

Model / Layout tabs

I also like to display the Model and Layout tabs.
Right click on the **Status Bar** as shown in #1 above. Select **Paper / Model**

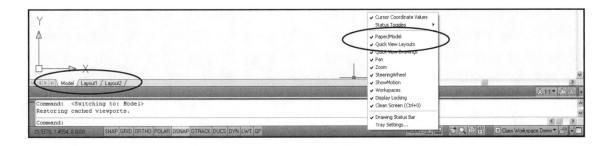

Notes:

LEARNING OBJECTIVES

After completing this lesson, you will be able to:

1. Create Template for use with Decimals.

LESSON 3

EXERCISE 3A

CREATE A MASTER DECIMAL SETUP TEMPLATE

The following instructions will guide you through creating a "Master" decimal setup template. This setup template is very similar to the setup template you may have made while using the Beginning Workbook but there are a few differences.

NEW SETTINGS

A. Begin your drawing with a different template as follows:
   1. Select the **NEW** command.
   2. Select template file **acad.dwt** from the list of templates.
   3. Select **Open**

B. Set the drawing **Units** as follows:
   1. Type **UNITS <enter>**
   2. Change the **Type** and **Precision** as shown
   3. Select **OK** button

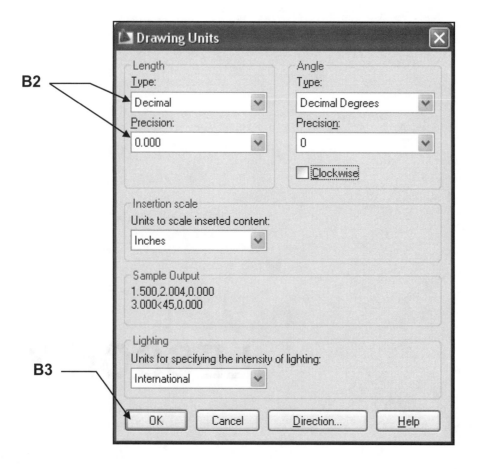

EXERCISE 3A continued

C. Set the **Drawing Limits** (Size of the drawing area) as follows:
   1. Type **Limits <enter>**
   2. Select **Off**

   "Off" means that your drawing area in Modelspace is unlimited.
   You will control the plotting with the layout tabs. (Paperspace)

D. Set the Grids and Snap as follows:
   1. Type **DS<enter>**
   2. Select the **Snap and Grid** tab
   3. Change the settings as shown below.
   4. Select the **OK** button.

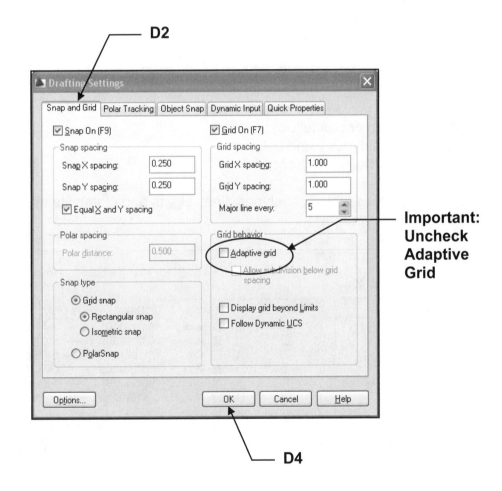

EXERCISE 3A continued

E. Change **Lineweight** settings as follows:
1. Right click on Lineweight button **(LWT)** on status line
2. Select **Settings**
3. Change to inches and adjust the Display scale as shown below.
4. Select the **OK** button.

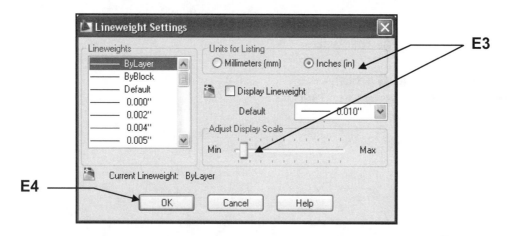

F. Create **new Layers** as follows:
1. Load the linetypes listed below.
   Center2, Hidden, Phantom2

2. Assign names, colors, linetypes, lineweights and plotability as shown below:

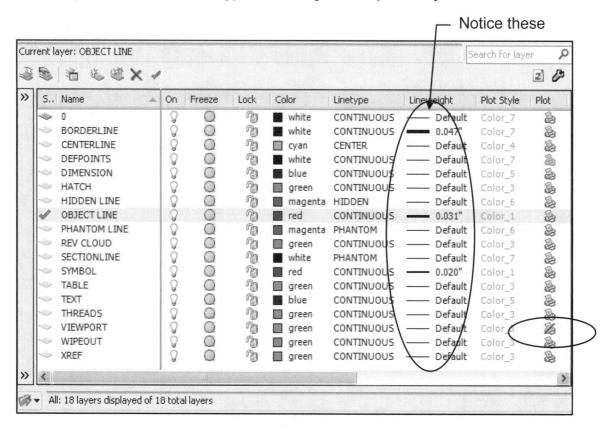

EXERCISE 3A continued

G. Create a new **Text Style** as follows:
1. Select **Text Style**
2. Create the text style **Text-Classic** using the settings shown below.
3. When complete, select **Set Current** and **Close.**

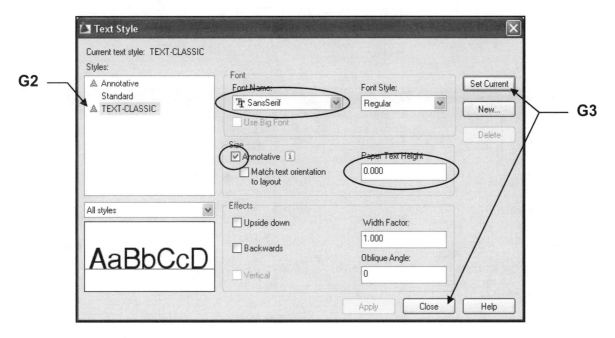

H. Create a **Dimension Style** as follows:
1. Select the **Dimension Style** command
2. Select the **New** button.

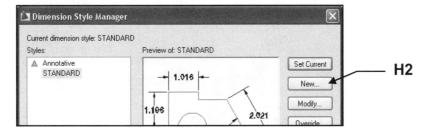

3. Enter **Name, Start with, Annotative, Use for** as shown below.
4. Select **Continue** button

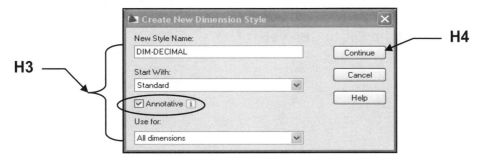

EXERCISE 3A continued

5. Select the **Primary Units** tab and change your setting to match the settings shown below.

5. Primary Units

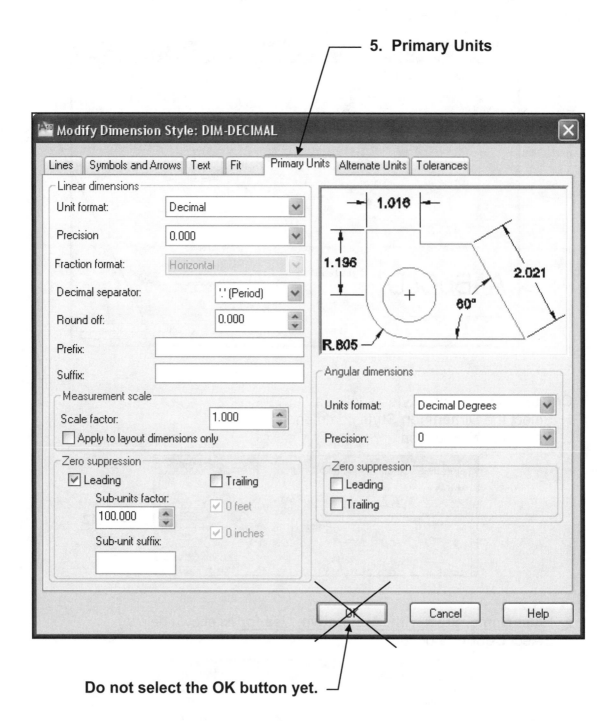

Do not select the OK button yet.

EXERCISE 3A continued

6. Select the **Lines** tab and change your settings to match the settings shown below.

6. Lines

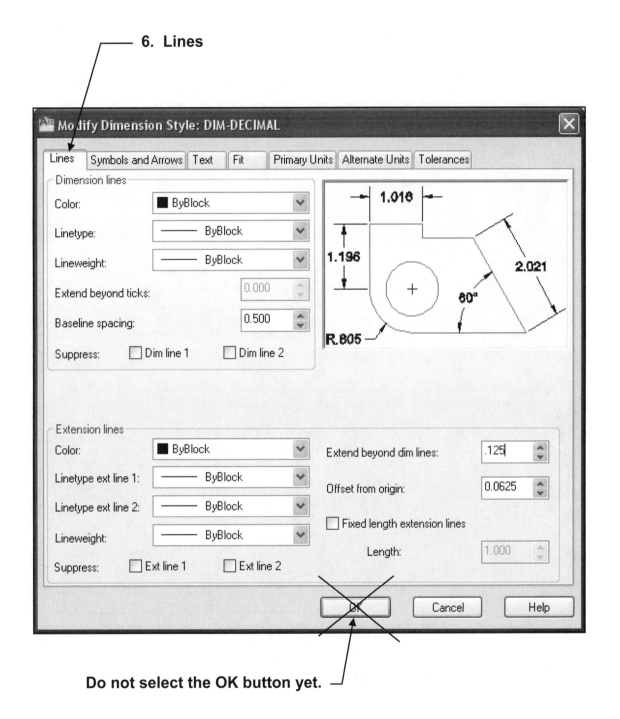

Do not select the OK button yet.

EXERCISE 3A continued

7. Select the **Symbols and Arrows** tab and change your setting to match the settings shown below.

7. Symbols and Arrows

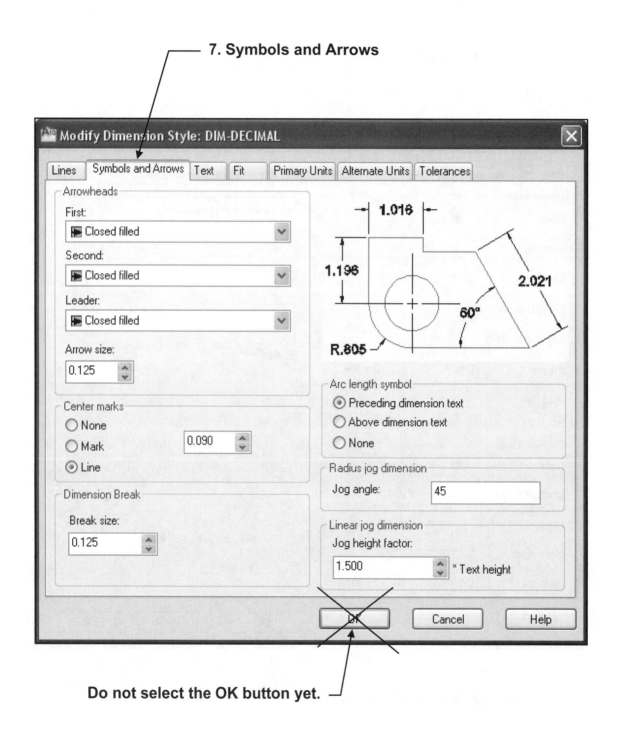

Do not select the OK button yet.

EXERCISE 3A continued

8. Select the **Text** tab and change your settings to match the settings shown below

8. Text

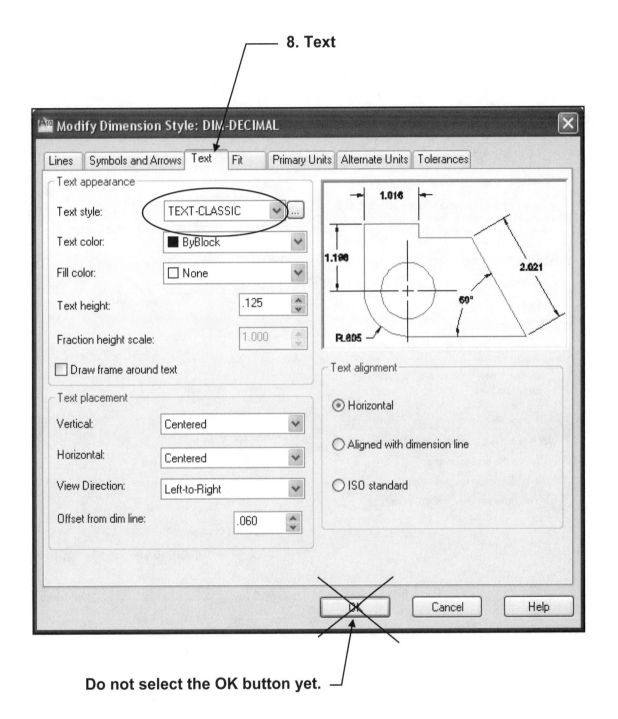

Do not select the OK button yet.

EXERCISE 3A continued

9. Select the **Fit** tab and change your setting to match the settings shown below.

10. Now select the **OK** button.

9. Fit

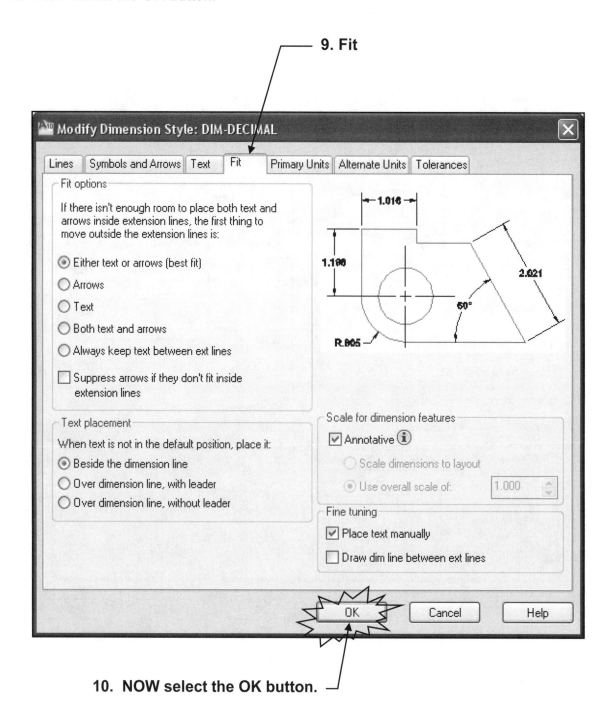

10. NOW select the OK button.

EXERCISE 3A continued

Your new style **"DIM-DECIMAL"** should be listed.

11. Select the **"Set Current"** button to make your new style "Class Style" the style that will be used.

12. Select the **Close** button.

13. **Important:** Save your drawing as **My Decimal Setup.**

11

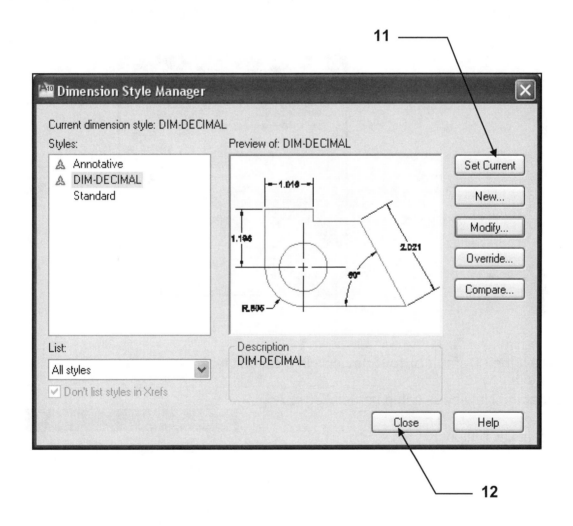

12

Continue to Exercise 3B

EXERCISE 3B

PAGE SETUP

Now you will select the printer and the paper size to use for printing.
You will use the Layout1 tab (Paper space).

A. Open **My Decimal Setup** (If not already open)

B. Select **Layout1** tab.

Refer to page 2-17 if you do not have these tabs.

Note: If the "Page Setup Manager" dialog box shown below does not appear automatically, right click on the Layout tab and select Page Setup Manager.

C. Select the **New...** button.

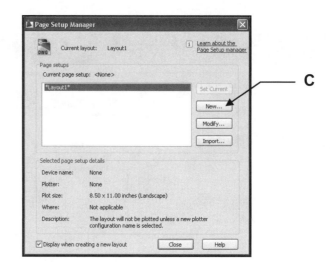

D. Select the **<Default output device>** in the **Start with**: list.

E. Enter the New page setup name: **Setup A**

F. Select **OK** button.

(I am assuming that your computer is attached to a printer. If not select Layout1)

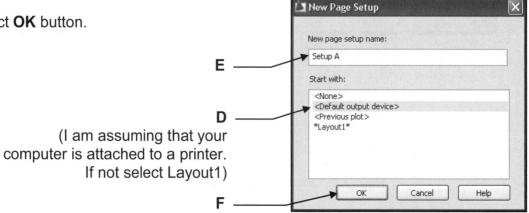

Continued on the next page...

EXERCISE 3B continued

This is where you will select the **print device**, **paper size** and the **plot offset**.

G. Select the **Printer / Plotter**
   Note: Your current system printer should already be displayed here. If you prefer another select the down arrow and select from the list. If the preferred printer is not in the list you must configure the printer. Refer to Appendix-A in Beg. Workbook.)

H. Select the **Paper Size**

I. Select **Plot Offset**

Notice the name you entered is now displayed as the page setup name.

G
(Yours will be different)

H
(Yours may be 8-1/2 x 11)

I
(Should stay at 0.00000)

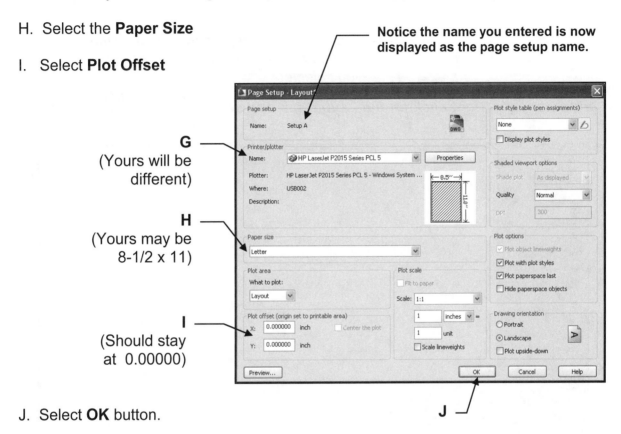

J. Select **OK** button.

J

K. Select **Setup A**.

L. Select the **Set Current** button.

M. Select the **Close** button.

L

K

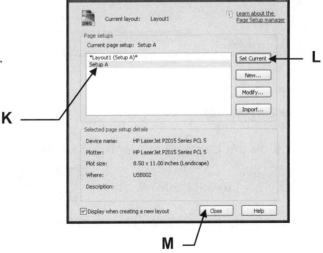

M

Continued on the next page...

EXERCISE 3B continued

You should now have a sheet of paper displayed on the screen.
This sheet is the size you specified in the "Page Setup".
This sheet is in front of Model Space.

The dashed line represents the maximum printing area for the printing device that you selected.
Any object outside of this area will not print.

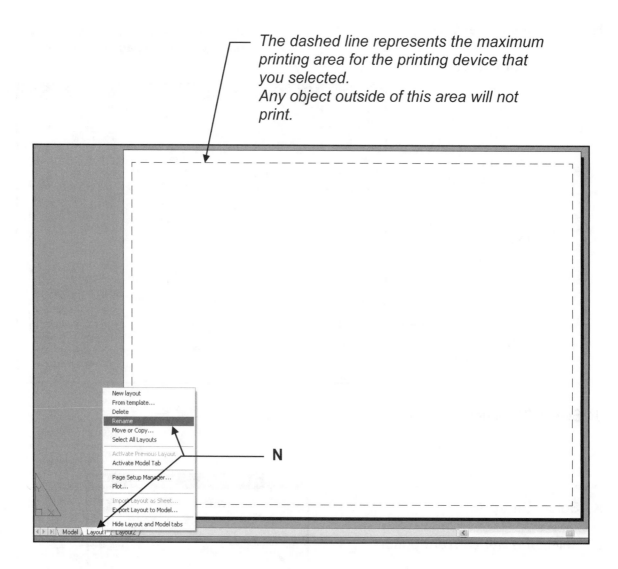

Rename the Layout tab

N. Right click on the **Layout1 tab** and select **Rename** from the list.

O. Enter the New Layout name **A Size**

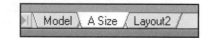

P. ***Very important:*** Save as **My Decimal Setup** again.

Continue on to **Exercise 3C**.....you are not done yet.

EXERCISE 3C

CREATE A BORDER AND TITLE BLOCK

A. Draw the border rectangle as large as you can within the dashed lines approximately as shown below using layer **Border**,

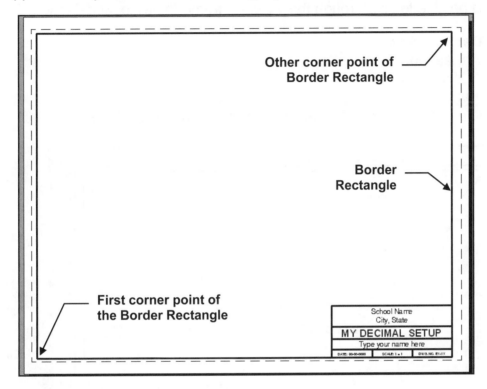

Other corner point of Border Rectangle

Border Rectangle

First corner point of the Border Rectangle

School Name
City, State
MY DECIMAL SETUP
Type your name here
DATE: 00-00-0000 SCALE: 1 = 1 DWG. NO. EX-XX

B. Draw the Title Block as shown using:
   1. Layers <u>Border</u> and <u>Text</u>.
   2. Multiline Text ; Justify <u>Middle Center</u> in each rectangular area.
   3. Text Style= Text-Classic   Text height = varies.

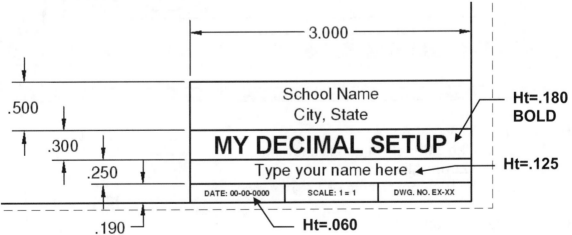

3.000

.500

.300

.250

.190

School Name
City, State

Ht=.180
BOLD

MY DECIMAL SETUP

Type your name here

Ht=.125

DATE: 00-00-0000   SCALE: 1 = 1   DWG. NO. EX-XX

Ht=.060

C. ***Very important:*** Save as **My Decimal Setup** again.

Continue on to **Exercise 3D**.....you are not done yet.

EXERCISE 3D

CREATE A VIEWPORT

The following instructions will guide you through creating a VIEWPORT in the Border
Layout sheet. Creating a viewport has the same effect as cutting a hole in the sheet of
paper. You will be able to see through the viewport frame (hole) to Modelspace.

A. Open **My Decimal Setup**
B. Select the **A Size** tab.
C. Select layer **Viewport**
D. Type: **MV <enter>**
E. Draw a Single viewport approximately as shown. (Turn off OSNAP)

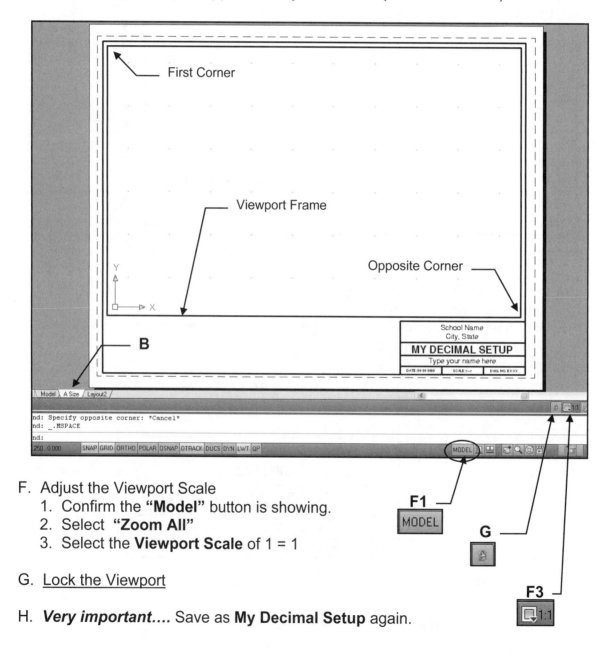

F. Adjust the Viewport Scale
    1. Confirm the **"Model"** button is showing.
    2. Select **"Zoom All"**
    3. Select the **Viewport Scale** of 1 = 1

G. Lock the Viewport

H. ***Very important….*** Save as **My Decimal Setup** again.

Continue on to **Exercise 3E**….you are not done yet.

3-16

EXERCISE 3E

PLOTTING FROM THE LAYOUT

The following instructions will guide you through the final steps for setting up the master template for plotting. These settings will stay with **My Decimal Setup** and you will be able to use it over and over again.

A. Open **My Decimal Setup**

B. Select the **A Size** layout tab.

   You should be seeing your Border and Title Block now.

C. Type **Plot <enter>**

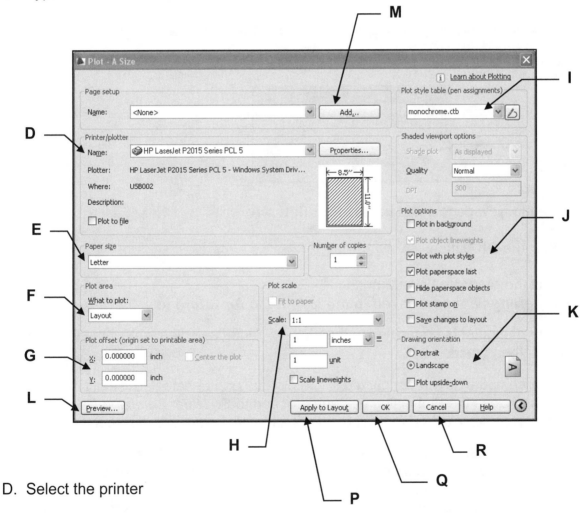

D. Select the printer

E. Select the Paper Size

F. Select the Plot Area

G. Plot offset should be 0.000000 for X and Y

Continued on the next page...

EXERCISE 3E continued

H. Select scale 1 : 1

I. Select the Plot Style Table **Monochrome.ctb**

J. Select the Plot options shown

K. Select Drawing Orientation: **Landscape**

L. Select **Preview** button.

>   If the drawing appears as you would like it, press the **Esc** key and continue
>   on to **M**.

>   If the drawing does not look correct, press the **Esc** key and check all your
>   settings, then preview again.

Note: The Viewport will not appear in the Preview because it is on a no plot layer.

M. Select the **ADD** button.

N. Type the New page setup name: **Plot Setup A**

O. Select the **OK** button.

P. Select the "Apply to Layout" button.
    The settings are now saved to the layout tab for future use.

Q. If your computer **is** connected to the plotter / printer selected, select the **OK** button
    to plot, then proceed to **S**.

R. If your computer is **not** connected to the plotter / printer selected, select the Cancel
    button to close the Plot dialog box and proceed to **S**. *Note: Selecting Cancel will
    cancel your selected setting if you did not save the page setup as specified in **M**.*

S. You are almost done. Now Save all of this work as a **Template.**
    1. Select **Application Menu / Save As / AutoCAD Drawing Template**
    2. Type: **My Decimal Setup.dwt**
    3. Select **Save** button.

>   Your **My Decimal Setup** master template is complete.

LEARNING OBJECTIVES

After completing this lesson, you will be able to:

1. Create master Architectural template

LESSON 4

EXERCISE 4A

CREATE A MASTER FEET-INCHES SETUP TEMPLATE

The following instructions will guide you through creating a "Master" Feet-inches setup template. This setup template is very similar to the setup template you may have made while using the Beginning workbook but there are a few differences.

NEW SETTINGS

A. Begin your drawing with a different template as follows:
    1. Select the **NEW** command.
    2. Select template file **acad.dwt** from the list of templates. (**Not** acad3D.dwt)
    3. Select **Open**

B. Set the drawing **Units** as follows:
    1. Type **UNITS <enter>**
    2. Change the **Type** and **Precision** as shown
    3. Select **OK** button

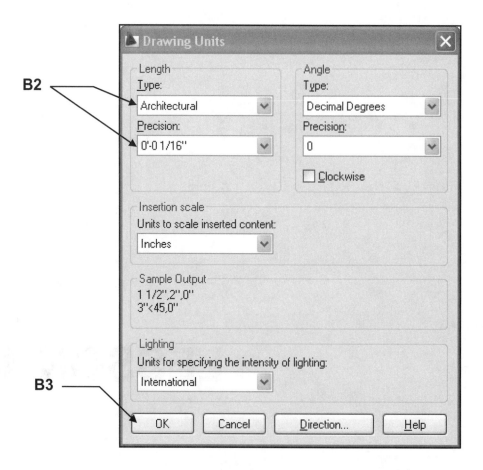

EXERCISE 4A continued

C. Set the **Drawing Limits** (Size of the drawing area) as follows:
   1. Type **Limits <enter>**
   2. Select **OFF**
   "Off" means that your drawing area in Modelspace is unlimited.
   You will control the plotting with the layout tabs. (Paperspace)

D. Set the Grids and Snap as follows:
   1. Type **DS<enter>**
   2. Select the **Snap and Grid** tab
   3. Change the settings as shown below.
   4. Select the **OK** button.

D2

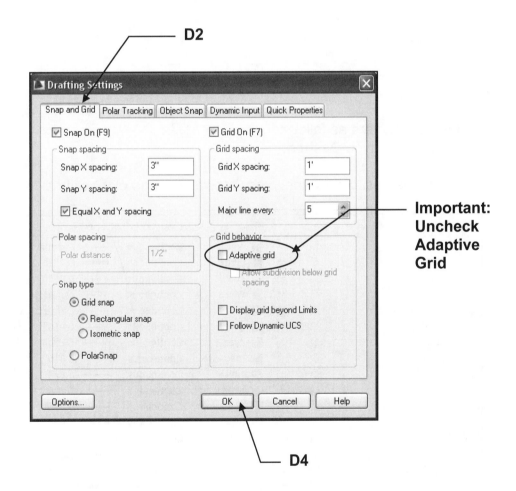

**Important:
Uncheck
Adaptive
Grid**

D4

Continued on the next page...

EXERCISE 4A continued

E. Change **Lineweight** settings as follows:
1. Right click on Lineweight button on status line
2. Select **Settings**
3. Change to inches and adjust the Display scale as shown below.
4. Select the **OK** button.

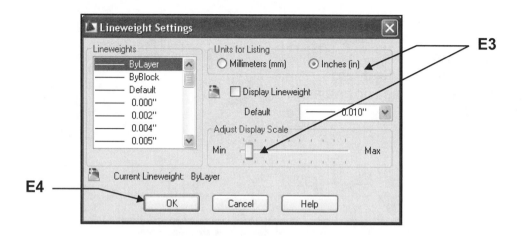

F. Create **new Layers** as follows:
1. Load the linetype **Dashed**.

2. Assign names, colors, linetypes, lineweights and plot ability as shown below:

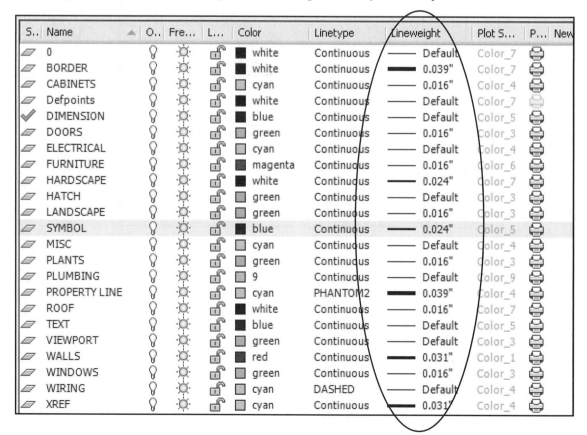

S..	Name	O..	Fre...	L...	Color	Linetype	Lineweight	Plot S...	P...	New
	0				white	Continuous	Default	Color_7		
	BORDER				white	Continuous	0.039"	Color_7		
	CABINETS				cyan	Continuous	0.016"	Color_4		
	Defpoints				white	Continuous	Default	Color_7		
✓	DIMENSION				blue	Continuous	Default	Color_5		
	DOORS				green	Continuous	0.016"	Color_3		
	ELECTRICAL				cyan	Continuous	Default	Color_4		
	FURNITURE				magenta	Continuous	0.016"	Color_6		
	HARDSCAPE				white	Continuous	0.024"	Color_7		
	HATCH				green	Continuous	Default	Color_3		
	LANDSCAPE				green	Continuous	0.016"	Color_3		
	SYMBOL				blue	Continuous	0.024"	Color_5		
	MISC				cyan	Continuous	Default	Color_4		
	PLANTS				green	Continuous	0.016"	Color_3		
	PLUMBING				9	Continuous	Default	Color_9		
	PROPERTY LINE				cyan	PHANTOM2	0.039"	Color_4		
	ROOF				white	Continuous	0.016"	Color_7		
	TEXT				blue	Continuous	Default	Color_5		
	VIEWPORT				green	Continuous	Default	Color_3		
	WALLS				red	Continuous	0.031"	Color_1		
	WINDOWS				green	Continuous	0.016"	Color_3		
	WIRING				cyan	DASHED	Default	Color_4		
	XREF				cyan	Continuous	0.031"	Color_4		

EXERCISE 4A continued

G. Create 2 new **Text Styles** as follows:
1. Select **Text Style**
2. Create the text style **Text-Classic** and **Text-Arch** using the settings shown below.

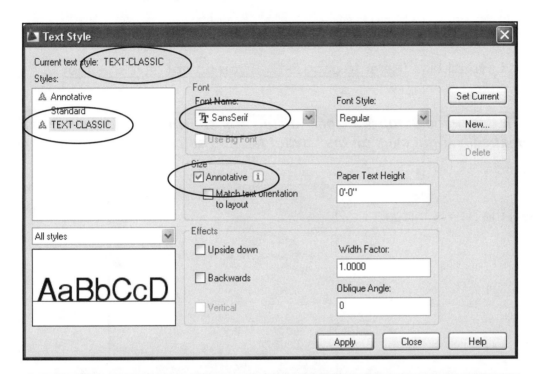

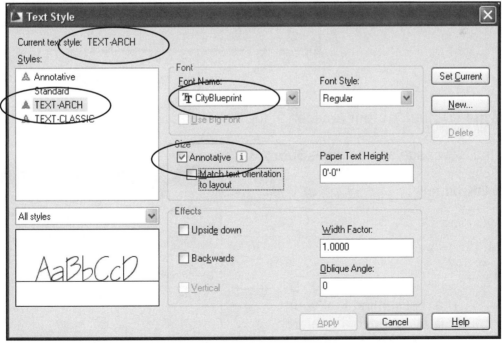

THIS NEXT STEP IS VERY IMPORTANT

H. **Save** all the settings you just created as follows:
1. Save as: **My Feet-Inches Setup**

EXERCISE 4B

PAGE SETUP

Now you will select the printer and the paper size to use for printing.
You will use the Layout1 tab (Paper space).

A. Open **My Feet-Inches Setup**

B. Select **Layout1** tab. **(Refer to page 2-17 if you do not have this tab)**

Note: If the "Page Setup Manager" dialog box shown below does not appear
automatically, right click on the Layout tab and select Page Setup Manager.

C. Select the **New...** button.

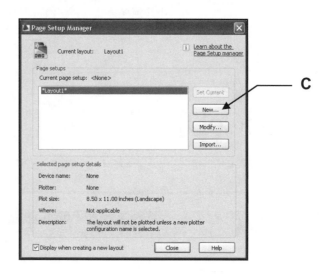

D. Select the **<Default output device>** in the Start with: list.

E. Enter the New page setup name: **Setup B**

F. Select **OK** button.

(I am assuming that your
computer is attached to a printer.
If not select Layout1)

EXERCISE 4B continued

This is where you will select the **print device**, **paper size** and the **plot offset**.

G. Select the **Printer / Plotter**
Note: Your current system printer should already be displayed here. If you prefer another select the down arrow and select from the list. If the preferred printer is not in the list you must configure the printer. Refer to Appendix-A in Beg. Workbook.)

H. Select the **Paper Size**

I. Select **Plot Offset**

Notice the name you entered is now displayed as the page setup name.

G
(Yours will be different)

H
(Yours may be 8-1/2 x 11)

I
(Should stay at 0.000000)

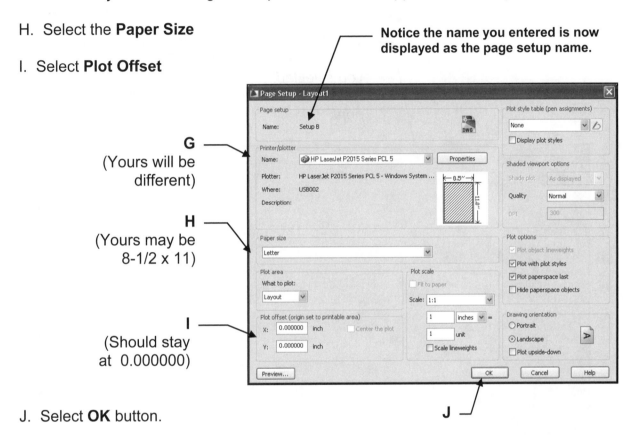

J. Select **OK** button.

J

K. Select **Setup B**.

L. Select the **Set Current** button.

M. Select the **Close** button.

K

L

M

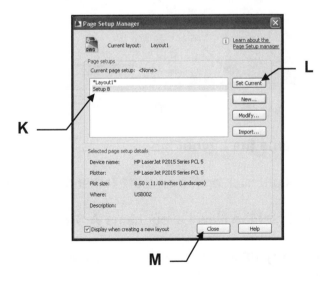

Continued on the next page...

4-7

EXERCISE 4B **continued**

You should now have a sheet of paper displayed on the screen.
This sheet is the size you specified in the "Page Setup".
This sheet is in front of Model Space.

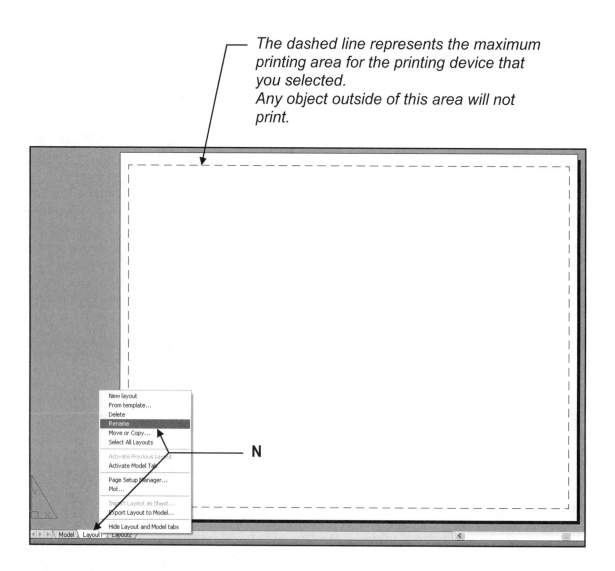

The dashed line represents the maximum printing area for the printing device that you selected.
Any object outside of this area will not print.

Rename the Layout tab

N. Right click on the Layout1 tab and select **Rename** from the list.

O. Enter the New Layout name **A Size**

P. *Very important:* Save as **My Feet-Inches Setup** again.

Continue on to **Exercise 4C**.....you are not done yet.

EXERCISE 4C

CREATE A BORDER AND TITLE BLOCK

A. Draw the <u>Border rectangle</u> as large as you can within the dashed lines approximately as shown below using <u>Layer Border</u>.

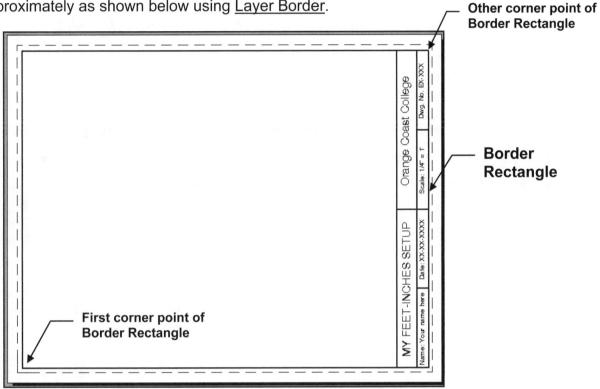

Other corner point of Border Rectangle

Border Rectangle

First corner point of Border Rectangle

B. Draw the Title Block as shown using:
1. Layers <u>Border</u> and <u>Text</u>.
2. **Single Line Text** ; Justify **Middle** in each rectangular area.
   (Hint: Draw diagonal lines to find the middle of each area, as shown)
3. Text Style= Text-Classic   Text height noted.

Note: You can draw the Title Block in place and add Text with a rotation angle of 90 or Draw the Title Block horizontal, add Text and then rotate it. You decide.

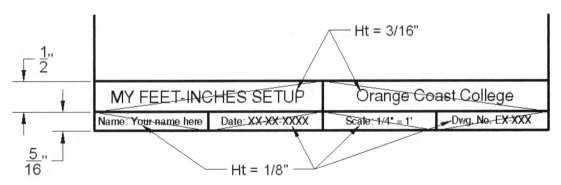

C. ***Very important:*** Save as **My Feet-Inches Setup** again.

Continue on to **Exercise 4D**…..you are not done yet.

EXERCISE 4D

CREATE A VIEWPORT

The following instructions will guide you through creating a VIEWPORT in the Border Layout sheet. Creating a viewport has the same effect as cutting a hole in the sheet of paper. You will be able to see through the viewport frame (hole) to Modelspace.

A. Open **My Feet-Inches Setup**
B. Select the **A Size** tab.
C. Select layer **Viewport**
D. Type: **MV <enter>**
E. Draw a Single viewport approximately as shown. (Turn off OSNAP temporarily)

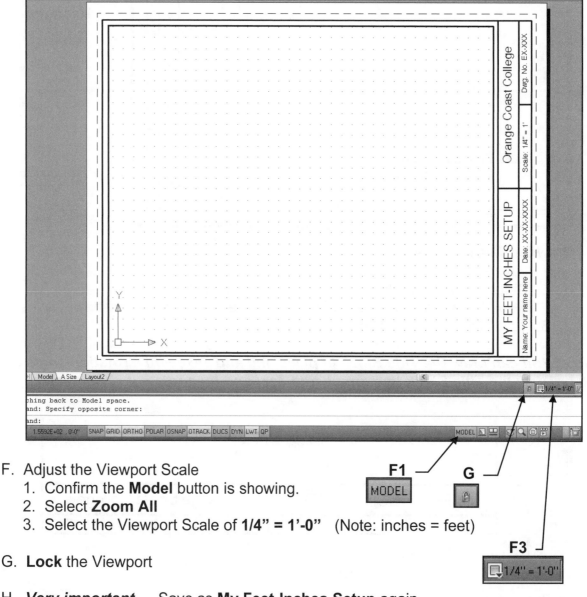

F. Adjust the Viewport Scale
   1. Confirm the **Model** button is showing.
   2. Select **Zoom All**
   3. Select the Viewport Scale of **1/4" = 1'-0"** (Note: inches = feet)

G. **Lock** the Viewport

H. ***Very important....*** Save as **My Feet-Inches Setup** again.

Continue on to **Exercise 4E**....you are not done yet.

EXERCISE 4E

PLOTTING FROM THE LAYOUT

The following instructions will guide you through the final steps for setting up the master template for plotting. These settings will stay with **My Feet-Inches Setup** and you will be able to use it over and over again.

In this exercise you will take a short cut by **importing** the **Plot Setup A** from **My Decimal Setup.dwt**. (Note: If the <u>Printer, Paper size and Plot Scale</u> is the same you can use the same Page Setup.)

> **Note: If you prefer <u>not</u> to <u>use</u> Import, you may go to 3-17 and follow the instructions for creating the Plot- page setup.**

A. Open **My Feet-Inches Setup**

B. Select the **A Size** layout tab.

   You should be seeing your Border and Title Block now.

C. Type **Plot <enter>**

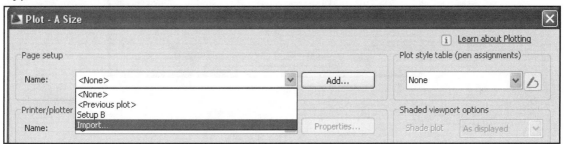

D. Select **Import...** from the Page Setup drop down menu

E. Find **My Decimal Setup.dwt** as follows:
   E1. Select Files of type: **Template [.dwt]**
   E2. Select **My Decimal Setup.dwt**
   E3. Select **OK** button.

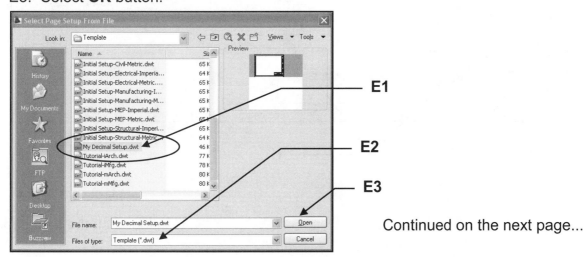

Continued on the next page...

EXERCISE 4E continued

F. Select **Plot Setup A** from the Page Setups list.

G. Select the **OK** button.

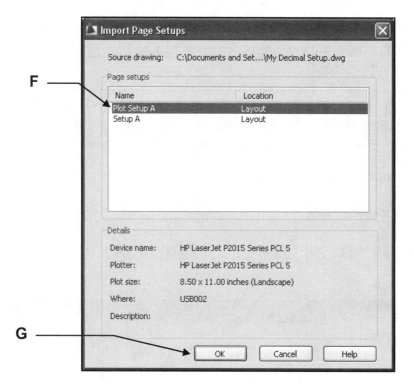

H. Select **Plot Setup A** from the Page Setup drop down list.
    (Note: It was not there before. You just imported it in from My Decimal Setup.dwt)

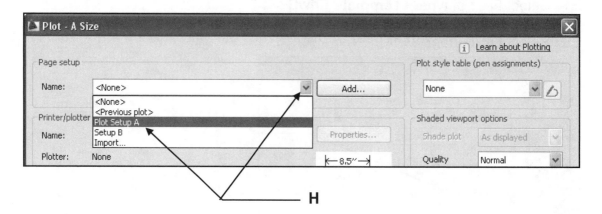

Continued on the next page...

EXERCISE 4E   continued

I. Review all the setting.

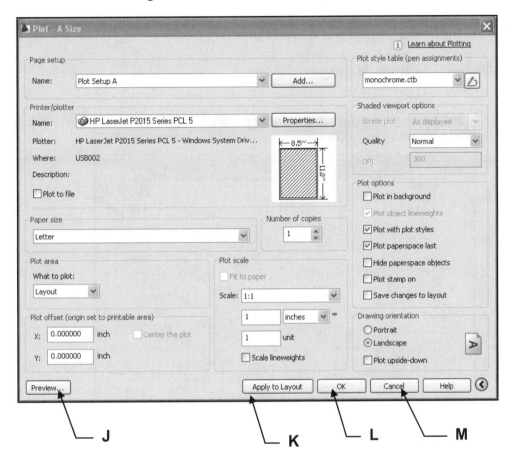

J. Select **Preview** button.

> If the drawing appears as you would like it, press the **Esc** key and continue
>
> If the drawing does not look correct, press the **Esc** key and check all your
> settings, then preview again.

Note: The Viewport will not appear in the Preview because it is on a no plot layer.

K. Select the **Apply to Layout** button.
(The settings are now saved to the layout tab for future use.)

L. If your computer **is** connected to the plotter / printer selected, select the **OK** button
to plot, then proceed to **N**.

M. If your computer is **not** connected to the plotter / printer selected, select the Cancel
button to close the Plot dialog box and proceed to **N**.

N. You are almost done. Now Save all of this work as a **Template.**
1. Select **File / Save As**
2. Save as: **My Feet-Inches Setup.dwt**

EXERCISE 4F

CREATE A NEW DIMENSION STYLE

1. Open **My Feet-Inches Setup** (If not already open)

2. Select the **Dimension Style Manager** command

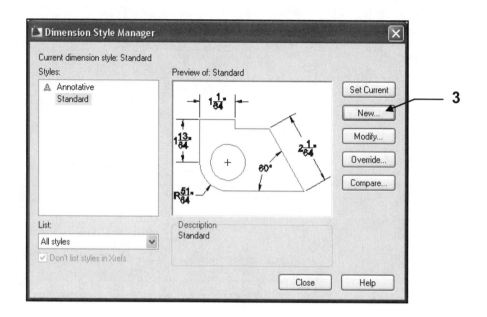

3. Select the **NEW** button.

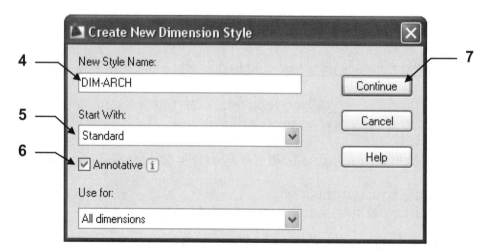

4. Enter **DIM-ARCH** in the "New Style Name" box.

5. Select **STANDARD** in the **"Start With:"** box.

6. Select **Annotative** box.

7. Select the **CONTINUE** button.

EXERCISE 4F continued

8. Select the **Primary Units** tab and change your settings to match the settings shown below.

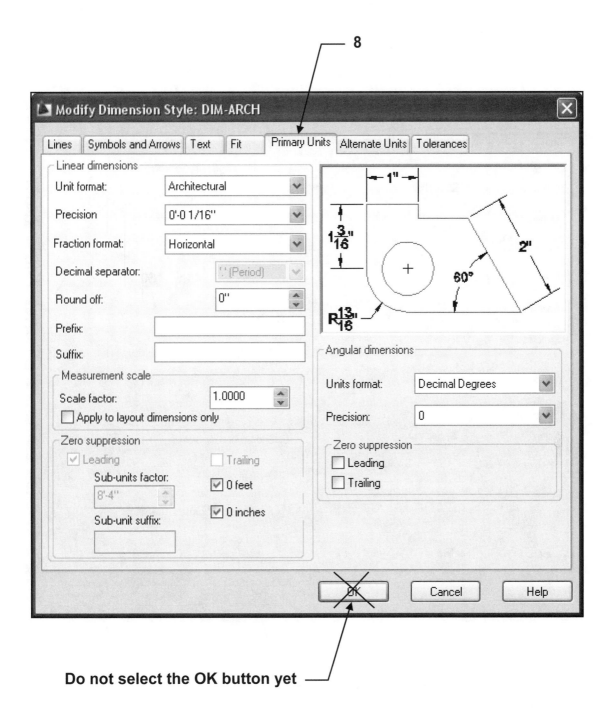

Do not select the OK button yet

EXERCISE 4F continued

9. Select the **Lines** tab and change your settings to match the settings shown below.

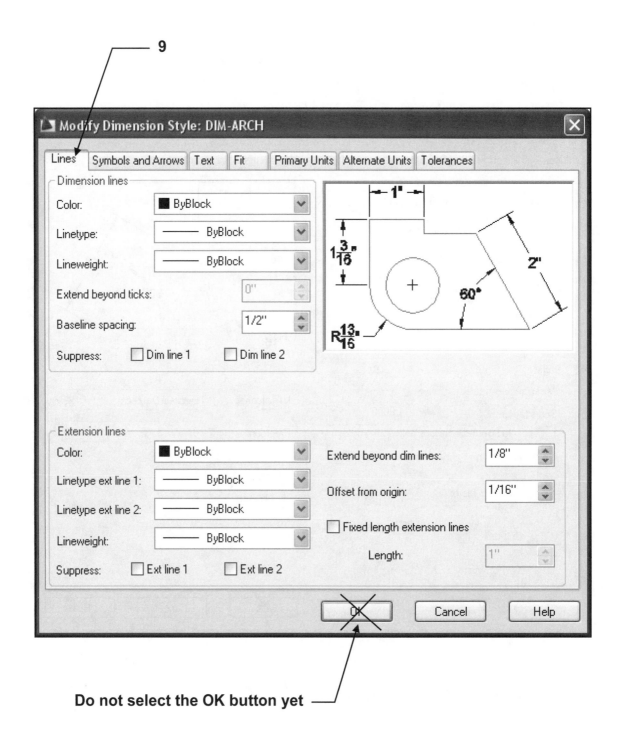

Do not select the OK button yet —

EXERCISE 4F continued

10. Select the **Symbols and Arrows** tab and change your settings to match the settings shown below.

10

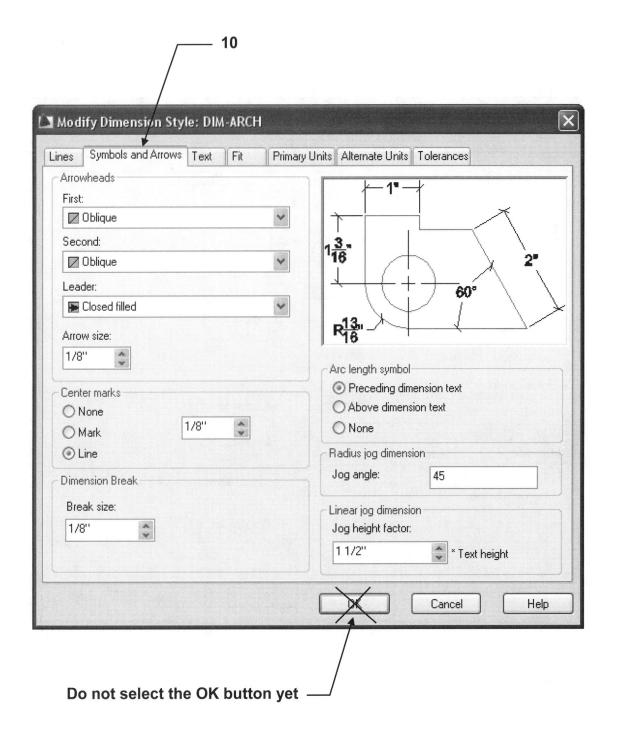

Do not select the OK button yet

EXERCISE 4F continued

11. Select the **Text** tab and change your settings to match the settings shown below.

11

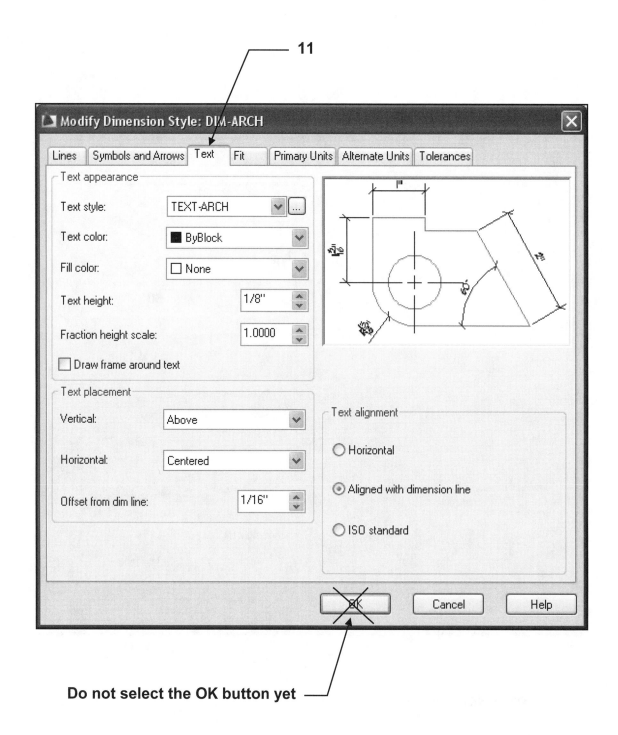

Do not select the OK button yet

EXERCISE 4F continued

12. Select the **Fit** tab and change your settings to match the settings shown below.

12

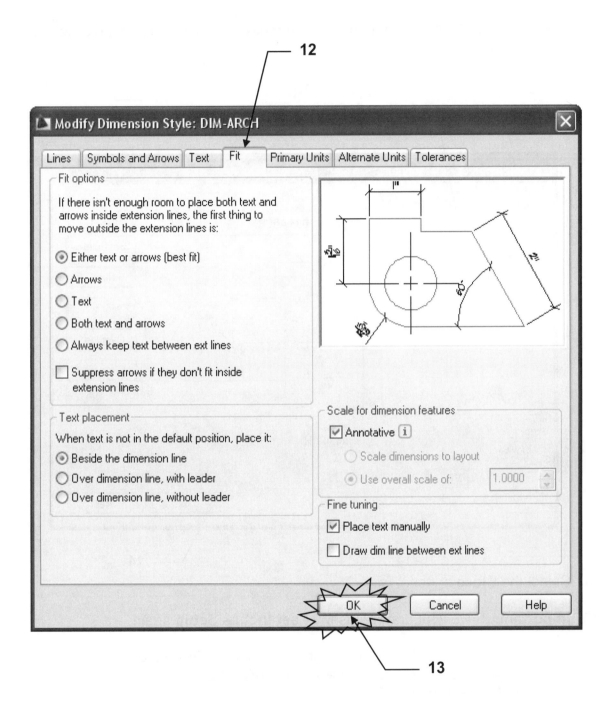

13. Now select the **OK** button.

EXERCISE 4F continued

Your new **DIM-ARCH** dimension style should now be in the list.

13. Select the **Set Current** button to make your new style **DIM-ARCH** the style that
     will be used.

New Style
listed here

13

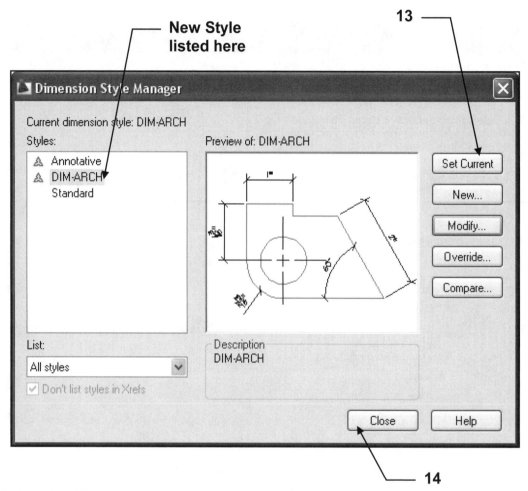

14. Select the **Close** button.

15. **Important:** Save your drawing as **My Feet-Inches Setup** again.

16. You are almost done. Now Save all of this work as a **Template.**
     a. Select **Application Menu / Save As / AutoCAD Drawing Template**
     b. Type: **My Feet-Inches Setup**
     c. Select **Save** button.

LEARNING OBJECTIVES

After completing this lesson, you will be able to:

1. Create a Table
2. Insert a Table
3. Modify an existing Table
4. Insert a Block into a Table Cell
5. Insert a Formula into a Table Cell
6. Create a Field
7. Update a Field
8. Edit an existing Field
9. Break a table

LESSON 5

TABLES

A Table is an object that contains data organized within columns and rows. AutoCAD's Table feature allows you to modify an existing Table Style or create your own Table Style and then enter text or even a block into the table cells. This is a very simple to use feature with <u>many</u> options.

This is an example of a Table.

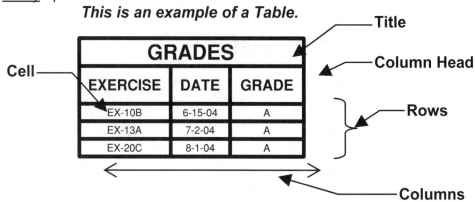

HOW TO CREATE A TABLE.

1. Select the Table Style command using one of the following:

   Ribbon = Annotate tab / Tables panel / ↘

   Keyboard = ts

   The following dialog box will appear.

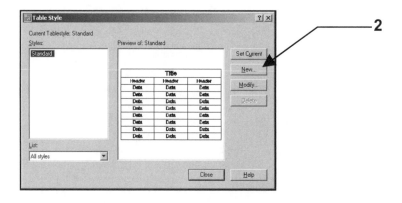

2. Select the **NEW** button. *The following dialog will appear.*

3. Enter the new Table Style name. *Note: When you create a new table style you always "Start With" an existing style and you specify the differences.*

4. Select the **Continue** button.

The New Table Style dialog box appears. Customize the table by selecting options described below.

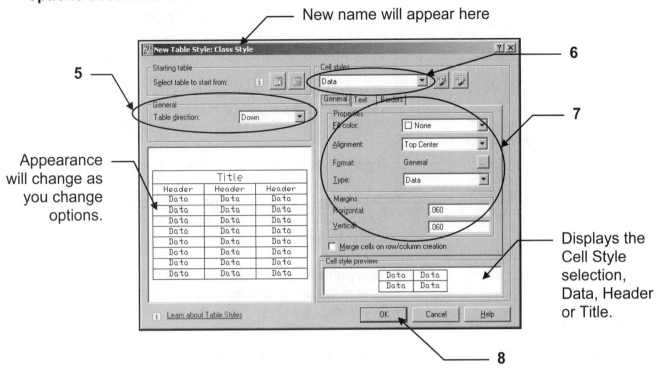

New name will appear here

5

Appearance will change as you change options.

6

7

Displays the Cell Style selection, Data, Header or Title.

8

5. Select which direction you want the table displayed.
   Up: Title and Header at the Bottom then data follows.
   Down: Title and Header at the Top then the data follows.

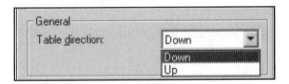

6. Select the Cell style that you wish to modify. (Data, Header or Title)

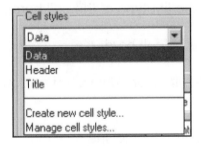

7. Select one of the <u>Properties tabs</u> after you have selected which "Cell Style" that you wish to modify.

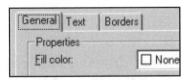

General tab

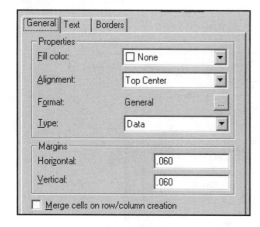

Fill Color: Background color of cell.

Alignment: Justification for the text inside the cell.

General button: Formats the cell content to: angle, currency, date, decimal number, general (default), Percentage, Point, Text or whole number.

Type: Cell style either Label or Data

Margins: Controls the spacing between the border of the cell and the cell content. Margins apply to all cells in the table.

Merge cells on row/column creation: Merges any new row or column created with the current cell style into one cell.

Text tab

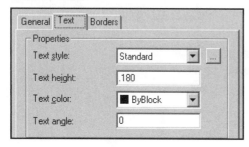

Text style: List all text styles in the drawing. Click the button to create a new text style.

Text height: Sets the text height

Text color: Specifies the text color

Text angle: Specify the rotation angle of the text.

Borders

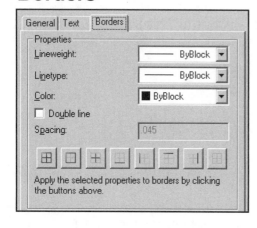

Lineweight: Selects lineweight for cell border

Linetype: Selects the linetype of the cell border.

Color: Select color for cell border.

Double line: Border lines are displayed as double lines.

Spacing: Specifies the double line spacing.

Border Buttons: Controls the appearance of the gridlines.

8. Select **OK** button after all selections have been made.

Note: Tables are saved in the drawing. If you open another drawing you need to copy the previously created table using Design Center. (Refer to Lesson 10)

9, Select the Table Style from the list of styles.

10. Select "**Set Current**" button.

11. Select the "**Close**" button.

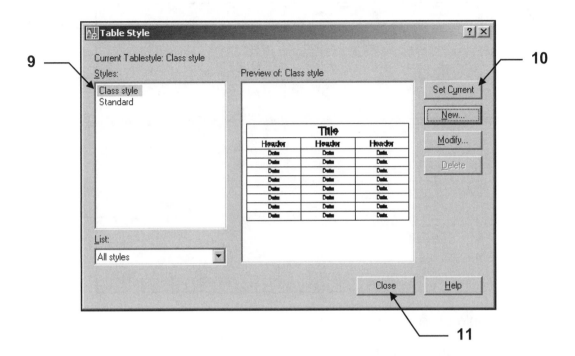

The new style is now displayed in the Tables Panel.

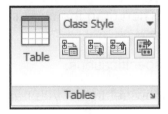

How to Insert a Table

1. Select **Table**  from the **Table panel**   or   type: **Table <enter>**

   The Insert Table dialog box will appear.

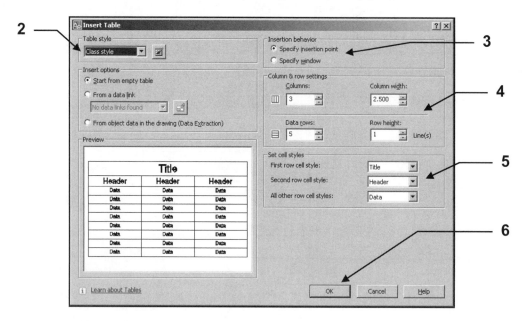

2. Select the Table Style

3. Select the Insertion behavior.

   Specify Insertion Point: When you select the OK button you select the location to insert the table with the Columns and Rows previously selected.

   Specify window: When you select the OK button you select the location for the upper left corner of the table. Then drag the cursor to specify the Column width and number of Rows, on the screen.

4. Specify the Column and Row specifications.

5. Set cell styles: Specifies the Row settings if you haven't selected a previously created Table Style.

6. Select **OK** button.

7. Place the Insertion Point. (The table will be attached to the cursor)

8. The Table is now on the screen waiting for you to fill in the data, header and title.

9. When you have filled all of the cells, select **Close**.

HOW TO INSERT A BLOCK INTO A TABLE CELL

1. Left click in the cell you wish to insert a block.

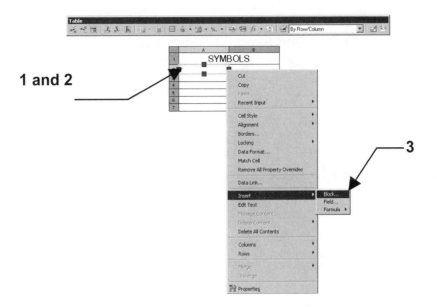

1 and 2

2. Right click to display the menu.

3. Select **Insert / Block** from the menu.

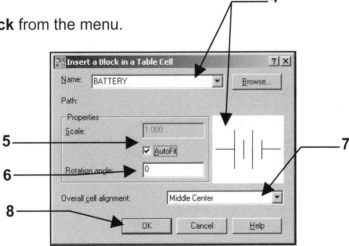

4. Select the **Block name** (a preview appears)

5. Select the **Scale.**
   Note: **AutoFit** will automatically size the block to fit within the cell.

6. Select **Rotation angle**.

7. Select the **Overall cell alignment**.

8. Select the **OK** button.

SYMBOLS	
⊣∣∣⊢	Battery
⊕▷⊣	Diode

HOW TO INSERT A FORMULA INTO A CELL

You may apply simple numerical operations such as Sum, Average, Count, set cells equal to other cells or even add an equation of your own.

The following examples are for Sum and Average operations.

SUM

1. Click in the Cell in which you want to enter a formula.

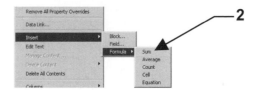

2. Right click and select:
   Insert / Formula / Sum

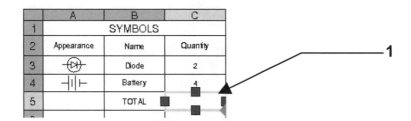

3. Select the cells content to sum using window.

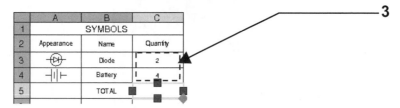

4. **Verify** the formula and select **Close Text Editor**.
   You may edit the formula if necessary.

Notice the formula disappears and the sum of the cells selected has been calculated. Also the value is shaded to make you aware that this cell has a formula in it.

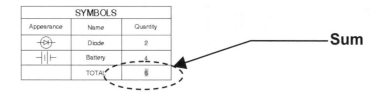

Sum

AVERAGE

1. Click in the Cell that you wish to enter a formula.

ROOM	TABLES	CHAIRS	COST
B14	2	4	100
F22	3	6	400
G7	4	8	500
TOTAL	9		
AVERAGE			

2

2. Set the data type, format and precision:
   a. Rt. Click in cell, **b**. Select **Data Format**
   c. Data Type: Decimal, **d**. Format: Decimal, **e**. Precision.

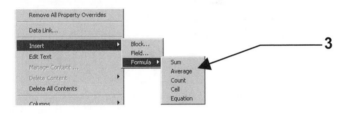

3. Right click and select:
   Insert Formula / Average

3

4. Select the cells content to average using a window.

ROOM	TABLES	CHAIRS	COST
B14	2	4	100
F22	3	6	400
G7	4	8	500
TOTAL	9		
AVERAGE			

4

5. Verify the formula and select **Close Text Editor**.
   You may edit the formula if necessary.

Notice the formula disappears and the average of the cells selected has been calculated.

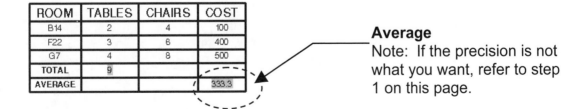

ROOM	TABLES	CHAIRS	COST
B14	2	4	100
F22	3	6	400
G7	4	8	500
TOTAL	9		
AVERAGE			333.3

Average
Note: If the precision is not what you want, refer to step 1 on this page.

HOW TO MODIFY AN EXISTING TABLE

1. Click once in a cell and the Table editing ribbon tools will appear.

   Experiment with all of the tools.

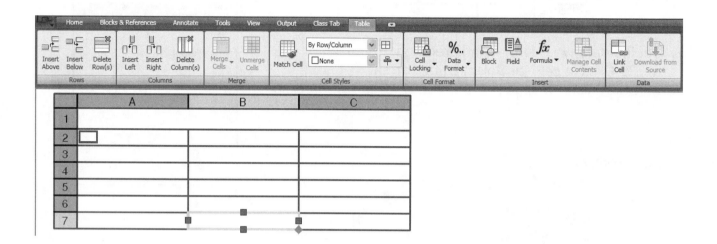

MODIFY A TABLE USING GRIPS

You may also modify tables using Grips. When editing with Grips, the left edge of the table remains stationary but the right edge can move. The upper left Grip is the Base Point for the table.

To use Grips, click on a table border line. The Grips should appear. Each Grip has a specific duty, shown below. To use a Grip, click on the Grip and it will change to red. Now click and drag it to the desired location.

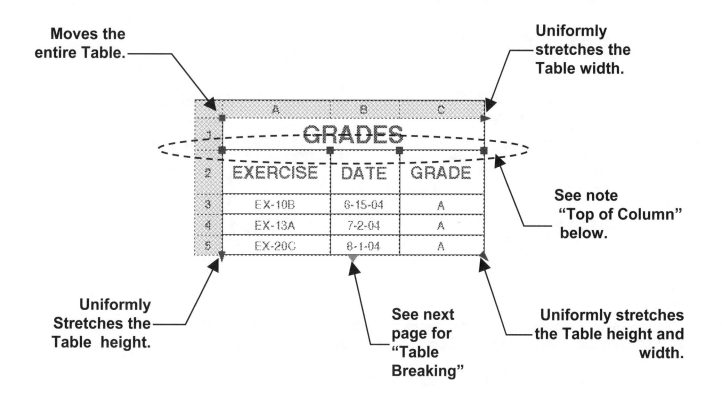

Top of Column
There is a Grip located at the top of each column line. These Grips adjust the width of the column <u>to the left</u> of the Grip.

The column will change but the width of the entire Table remains unchanged.

If you hold the CTRL key down while moving a column Grip, the entire Table adjusts at the same time.

AUTOFILL GRIP

The AutoFill grip allows you to fill the cells automatically by increments of 1 by selecting 1 cell and dragging to the next cell.
Make sure the cell data format is set correctly. (Refer to 5-13)

1. Place data in a cell. (**Important:** Data format should be "whole number". Refer to 5-13)

2. Click on the AutoFill Grip.

3. Drag the cursor down to the desired cells.

4. Cell data advances by 1 unit.

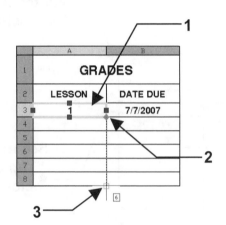

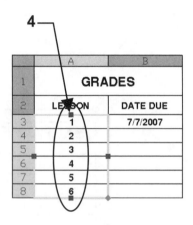

If you select **2 cells** AutoFill will follow the incremental advance between the 2 cells.
For example, if the date advances by 1 week rather than 1 day.

1. Fill 2 cells. (**Important:** Set data format to "date" Refer to 5-13)

2. Select the AutoFill Grip

3. Drag the cursor down to the desired cells.

4. Cell data advances by 1 week.

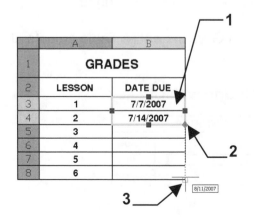

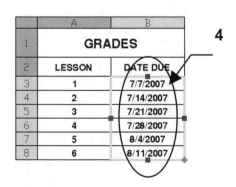

AUTOFILL OPTIONS

You may specify how you want the cells formatted.

1. Left click on the AutoFill Grip.

2. Right click on the AutoFill Grip.

3. Select an Option from the List.   (Description of each is shown below.)

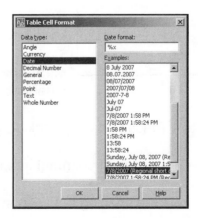

o   **Fill Series**. Fills the subsequent cells with the data from the selected cell and advances at a rate of 1 unit. The format of the selected cell is maintained.

o   **Fill Series Without Formatting**. Same as Fill Series except the formatting is not the same as the selected cell.

o   **Copy Cells**. Duplicates the values and the formatting of the selected cells.

o   **Copy Cells Without Formatting**. Same as Copy Cells except the formatting is not the same as the selected cell.

o   **Fill Formatting Only**. Fills only cell formatting for the selected cells. Cell values are ignored.

To change the data formatting within a cell.

1. Left click inside the cell.

2. Right click and select **Data Format....**

3. Select from the Table Cell Format .

TABLE BREAKING

A long table can be broken into 2 separate tables using the Table Breaking Grip.

1. Click on the table border line.

2. Click on the Table Breaking Grip.

3. Move the cursor **up** to the cell location for the Table Break and click.

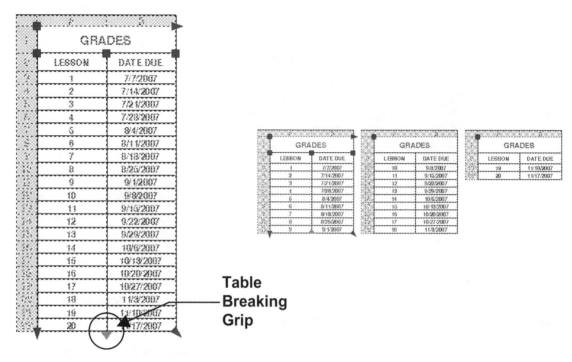

Table Breaking Grip

To make changes to the broken table
1. Right click on the Table and select **Properties**.
2. Scroll down to **Table Breaks**

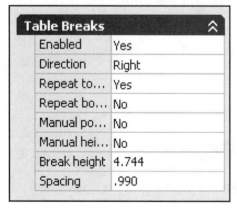

Enabled – Allows Table Breaking or Not

Direction – Places the broken table to the left, right or down from the original table.

Repeat Top Labels – Repeats the Top labels or not

Repeat Bottom Labels – Repeats the Bottom labels if your table has labels on the bottom or not.

Manual Positions – Allows you to move the segments independently using the Move Grip on each.

Manual Heights – Places a table breaking grip on each segment so each may be additionally broken

Break Height – Breaks each segment into equal heights except possibly the last.

Spacing – Controls the spacing between segments.

FIELDS   (Not available in LT version)

A **Field** is a string of text that has been set up to display data that it gets from another source. For example, you may create a field that will display the Circumference of a specific Circle within your drawing. If you changed the diameter of that Circle you could "update" the field and it would display the new Circumference.

Fields can be used in many different ways. After you have read through the example below you will understand the steps required to create and update a Field. Then you should experiment with some of the other Field Categories to see if they would be useful to you. Consider adding Fields to a cell within a Table.

CREATE A FIELD

1. Draw a 2" Diameter Circle and place it anywhere in the drawing area.
2. Select "**Insert** tab / **Field**

5. Select the Object Type button.

3. Select the Field Category: Objects

4. Select the Field Name: Object

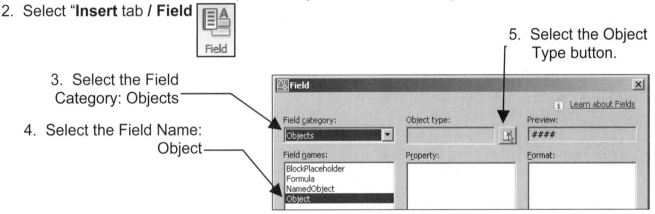

6. Select the Circle that you just drew.

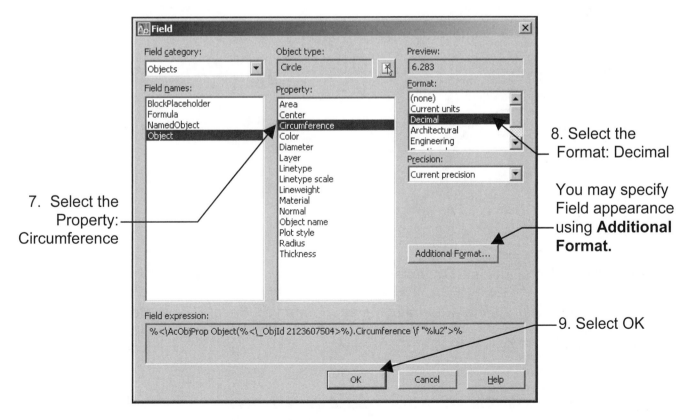

7. Select the Property: Circumference

8. Select the Format: Decimal

You may specify Field appearance using **Additional Format.**

Field expression:
%<\AcObjProp Object(%<_ObjId 2123607504>%).Circumference \f "%lu2">%

9. Select OK

10. Place the "Field" inside the Circle.

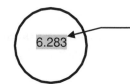

Note: Watch the command line closely. You may change the "Height" and "Justification" of the text before you place it.

Notice that the Field appears with a gray background. This background will not plot. The background can be turned off but I find it helpful to be able to visually distinguish a Field from plain text. If you wish to turn it off use: Options / User Preferences tab. In the Fields section, uncheck the "Display background of Fields" box.

UPDATE A FIELD

Now let's change the diameter of that Circle and see what happens to the Field.

1. Change the size of the Circle.
    a. Select the Circle.
    b. Right click and select **Properties.**
    c. Change the Diameter to 4.
    d. Close the Properties Palette.
    e. Press ESC key to clear the Grips.

Notice that the Field has not changed yet.
In order to see the new Circumference value, you must "Update the Field".

2. Type **Regen <enter>.**

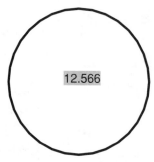

The Field updated to the new Circumference value.

Note: The Field will update automatically each time you Save, Plot or Regenerate the drawing.

Editing FIELDS

Editing a Field is very easy. The process is basically the same as creating a Field.

EDIT A FIELD

1. Double click the **Field text**. *The Multiline Text Editor will open*.
2. Right click on the **Field text**.
3. Select **Edit Field** from the menu.
4. Make the changes.
5. Select **OK**.
6. You may also add more text to the Field text but you may not change the text within the Field.
   Example: Add the words "Circle Circumference" under the Field text on the previous pages.
7. Select **OK** to exit the Multiline Text Editor.

ADD A FIELD TO A TABLE CELL

1. Select the cell.
2. Right click.
3. Select "**Insert / Field**" from the menu.
4. Create a Field as described on page 5-15.

Questions about Fields

1. What happens when you Explode a Field?

The Field will convert to normal text and will no longer update.

EXERCISE 5A

CREATE A NEW TABLE STYLE

A. Open **My Decimal Setup.**

B. Create a New Table.

    1. Type **TS** <enter>   (Refer to page 5-2)

    2. Select the **New** button.

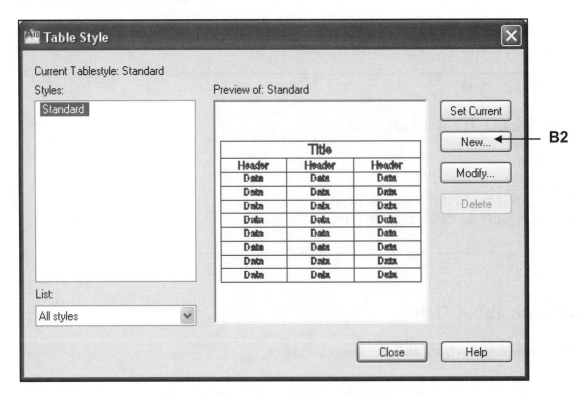

    3. Enter new table style "name" and start with "Standard"

    4. Select **Continue** button.

Continued on the next page...

EXERCISE 5A continued

C. Enter **"Title" Cell Type** settings.

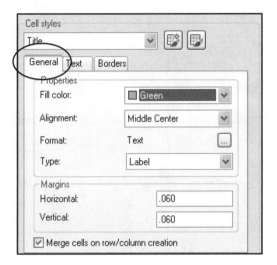

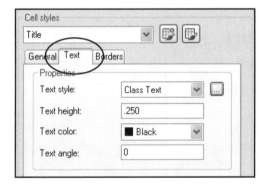

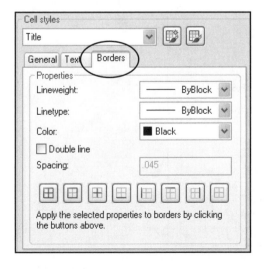

Continued on the next page...

EXERCISE 5A continued

C. Enter **"Header" Cell Type** settings.

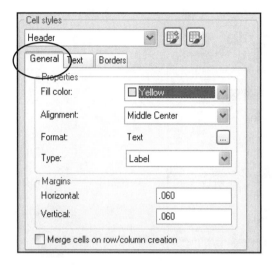

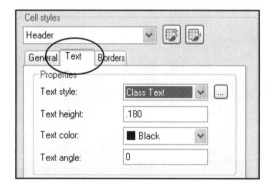

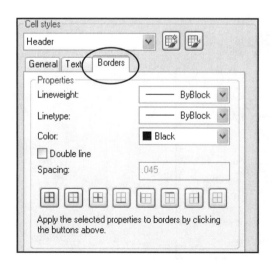

Continued on the next page...

EXERCISE 5A   continued

C. Enter **"Data" Cell Type** settings.

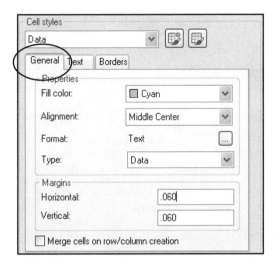

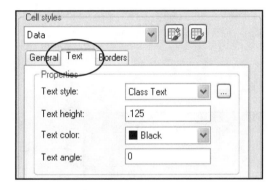

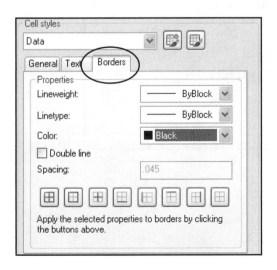

Continued on the next page...

EXERCISE 5A continued

D. Select the **OK** button.

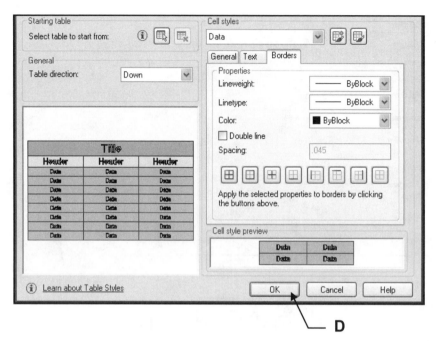

E. Set the new table style current.

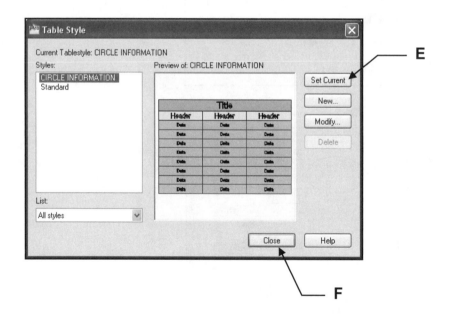

F. Select **Close** button.

G. Save as: **EX-5A**

EXERCISE 5B

INSERT A TABLE

1. Open **EX-5A**

2. Insert the Table shown below using the following Column and Width settings.

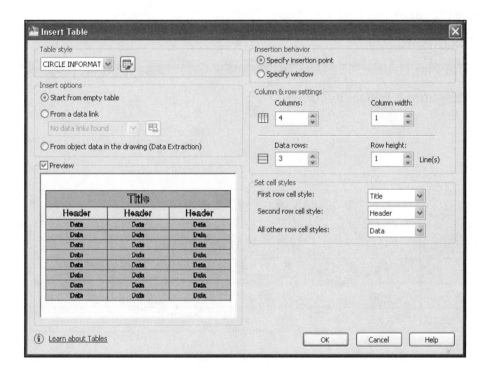

3. Enter the **Title** and **Header** text.

CIRCLE INFORMATION			
ITEM	DIA	CIR	AREA

4. Save as: **EX-5B**

EXERCISE 5C

MODIFY AN EXISTING TABLE

1. Open **EX-5B**

2. Modify the **Column Header** "Cir" to "Circumference".

   a. First change the Cell width to 2.750

   b. Edit the text

3. Add the data for column "ITEM" (Try "AutoFill Grip" described on page 5-12

4. Add the diagonal guidelines if desired.

CIRCLE INFORMATION			
ITEM	DIA	CIRCUMFERENCE	AREA
1			
2			
3			

5. Save as: **EX-5C**

EXERCISE 5D

ADD FIELDS TO AN EXISTING TABLE

1. Open **EX-5C**

2. Draw 3 Circles and place their item number (.25 Ht) in the middle, as shown below.

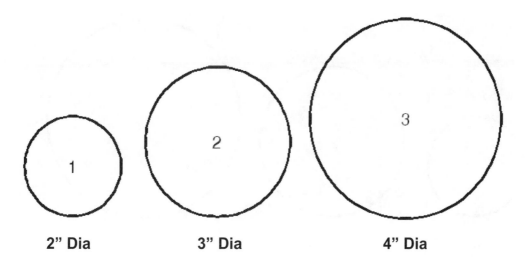

    2" Dia   **3" Dia**   **4" Dia**

3. Add the <u>FIELDS</u> for **DIA, CIRCUMFERENCE** and **AREA** in the appropriate Data Cells within the "Circle Information" Table shown below.

Text Ht = .125   Justification = Middle Center

CIRCLE INFORMATION			
ITEM	DIA	CIRCUMFERENCE	AREA
1	2.000	6.283	3.142
2	3.000	9.425	7.069
3	4.000	12.566	12.566

4. Save as: **EX-5D**

EXERCISE 5E

UPDATE A FIELD

1. Open **EX-5D**

2. Modify the Diameters of the 3 Circles using **Properties**.

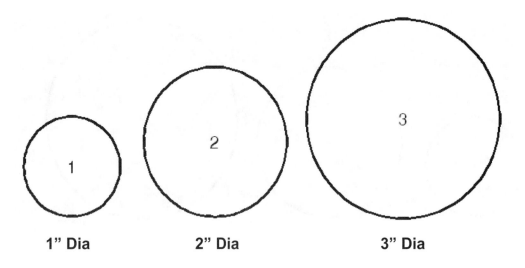

| 1" Dia | 2" Dia | 3" Dia |

3. Update the Fields: Type: **Regen** <enter> (Refer to page 5-16)

CIRCLE INFORMATION			
ITEM	DIA	CIRCUMFERENCE	AREA
1	1.000	3.142	0.785
2	2.000	6.283	3.142
3	3.000	9.425	7.069

4. Save as: **EX-5E**

EXERCISE 5F

USING AUTOFILL GRIP

1. Open **EX-5E**

   You are using 5E because it has the "Circle Information" table style saved in it.

2. Draw the table shown below using table style "Circle Informtion".

3. Column width = 1.50

4. Try **AutoFill Grip** to fill the cells. It is really not difficult and may save you time.

   (Refer to page 5-12)

GRADES	
LESSON 1	DUE DATE
1	7/7/2007
2	7/14/2007
3	7/21/2007
4	7/28/2007
5	8/4/2007
6	8/11/2007
7	8/18/2007
8	8/25/2007
9	9/1/2007
10	9/8/2007
11	9/15/2007
12	9/22/2007

5. Save as: **EX-5F**

EXERCISE 5G

BREAKING A TABLE

1. Open **EX-5F**

2. Break the table as shown below. (Refer to page 5-14)

GRADES	
LESSON 1	DUE DATE
1	7/7/2007
2	7/14/2007
3	7/21/2007
4	7/28/2007
5	8/4/2007

6	8/11/2007
7	8/18/2007
8	8/25/2007
9	9/1/2007
10	9/8/2007
11	9/15/2007
12	9/22/2007

3. Now change the broken table to **"Down"**.

GRADES	
LESSON 1	DUE DATE
1	7/7/2007
2	7/14/2007
3	7/21/2007
4	7/28/2007
5	8/4/2007

6	8/11/2007
7	8/18/2007
8	8/25/2007
9	9/1/2007
10	9/8/2007
11	9/15/2007
12	9/22/2007

4. Save as: **EX-5G**

LEARNING OBJECTIVES

After completing this lesson, you will be able to:

1. Create an Isometric object.
2. Create an Ellipse on an Isometric plane.

LESSON 6

ISOMETRIC DRAWINGS

An ISOMETRIC drawing is a pictorial drawing. It is primarily used to aid in visualizing an object. There are 3 faces shown in one view. They are: Top, Right and Left. AutoCAD provides you with some help constructing an isometric drawing but you will be doing most of the work. The following is a description of the isometric assistance that AutoCAD provides. **Note: This is not 3D. Refer the "Introduction to 3D" section in this workbook.**

ISOMETRIC SNAP and GRID

1. Select **DRAFTING SETTINGS** by typing **DS <enter>** or right click on the Grid Status bar button then select **Settings** from the short cut menu.

   The following dialog box should appear.

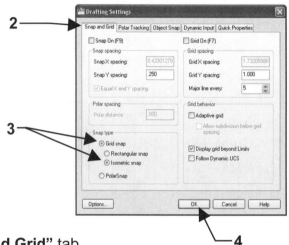

2. Select the "**Snap and Grid**" tab.
3. Select "**Grid snap**" and "**Isometric snap**".
4. Select **OK.**

NOTE: The Grid pattern and Cursor should have changed. The grids are on a 30 degree angle. The cursor's horizontal crosshair is now on a 30 degree angle also. This indicates that the LEFT ISOPLANE is displayed.

ISOPLANES

There are 3 isoplanes, Left, Top and Right. You can toggle to each one by pressing the **F5** key. When drawing the left side of an object you should display the Left Isoplane. When drawing the Top of an object you should display the Top Isoplane. When drawing the Right side of an object you should display the Right Isoplane.

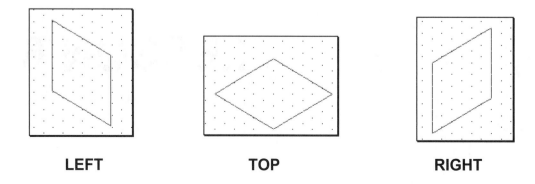

| LEFT | TOP | RIGHT |

5. **Grid, Snap and Ortho should be ON**.

6. Now try drawing the 3 x 3 x 3 cube below using the "Line" command.
   a. Start with the Left Isoplane. (You may use **DDE** but also use Ortho)
   b. Next change to the Top Isoplane (Press F5 once) and draw the top.
   c. Next change to the Right Isoplane (Press F5 once) and draw the Right side.

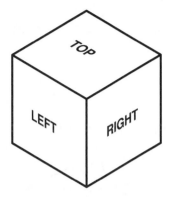

ISOMETRIC ELLIPSE

An isometric ellipse is a very helpful option when drawing an isometric drawing. AutoCAD calls an isometric ellipse an **ISOCIRCLE**. This option is located in the Ellipse command. ***IMPORTANT: the ISOCIRCLE option can only appears if you have selected the <u>Isometric Grid and Snap</u>. This option will not appear if you select the <u>Rectangular Grid and Snap</u>.***

1. Change to the Left Isoplane. (Isometric Grid and Snap must be on.)
2. Select the **Ellipse** "<u>**Axis End**</u>" command.   ***Do not select "Ellipse, Center")***

   Command: _ellipse
   Specify axis endpoint of ellipse or [Arc/Center/Isocircle]: ***type "I" <enter>***
   Specify center of isocircle: ***specify the center location for the new isocircle***
   Specify radius of isocircle or [Diameter]: ***type the radius <enter>***

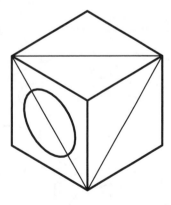

3. Now try drawing an Isocircle on the top and right, using the method above. Don't forget to change the "Isoplane" depending on where you draw the Isocircle.

EXERCISE 6A

ISOMETRIC ASSEMBLY

1. Start a **New** file using **My Decimal Setup.dwt**.
2. Select the **Model** tab
3. Set the **"Snap type and Style"** to isometric snap (Refer to page 6-2)
4. Change to isoplane **"Top"**
5. Draw the objects below. Do not dimension
6. Save as: **EX-6A**

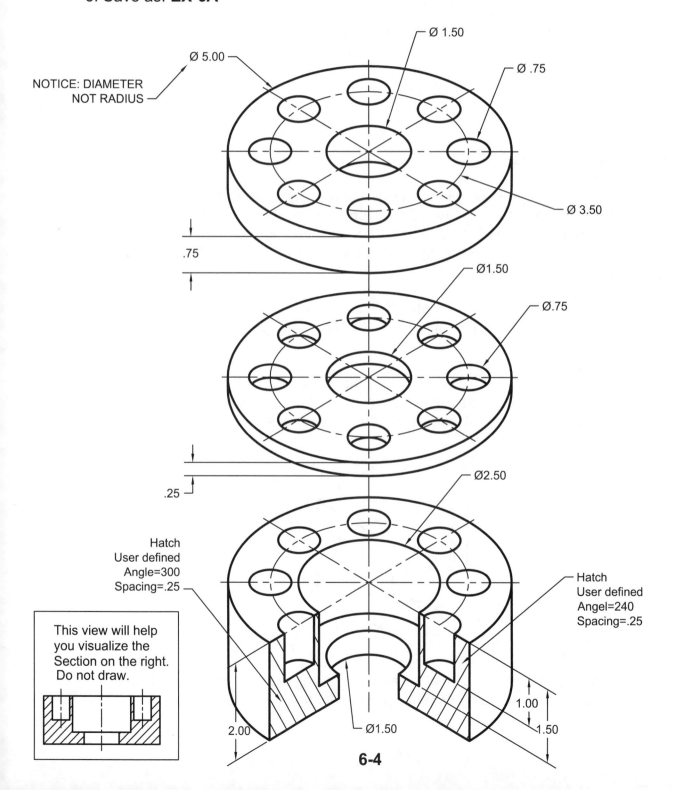

NOTICE: DIAMETER
NOT RADIUS

Ø 5.00

Ø 1.50

Ø .75

Ø 3.50

.75

Ø1.50

Ø.75

.25

Ø2.50

Hatch
User defined
Angle=300
Spacing=.25

Hatch
User defined
Angel=240
Spacing=.25

This view will help
you visualize the
Section on the right.
Do not draw.

2.00

Ø1.50

1.00

1.50

6-4

EXERCISE 6B

ISOMETRIC OBJECT

1. Start a **New** file using **My Decimal Setup.dwt**.
2. Select the **Model** tab
3. Set the **"Snap type and Style"** to isometric snap (Refer to page 6-2)
4. Change to isoplane with F5 when necessary.
5. Draw the objects below. Do not dimension
6. Save as: **EX-6B**

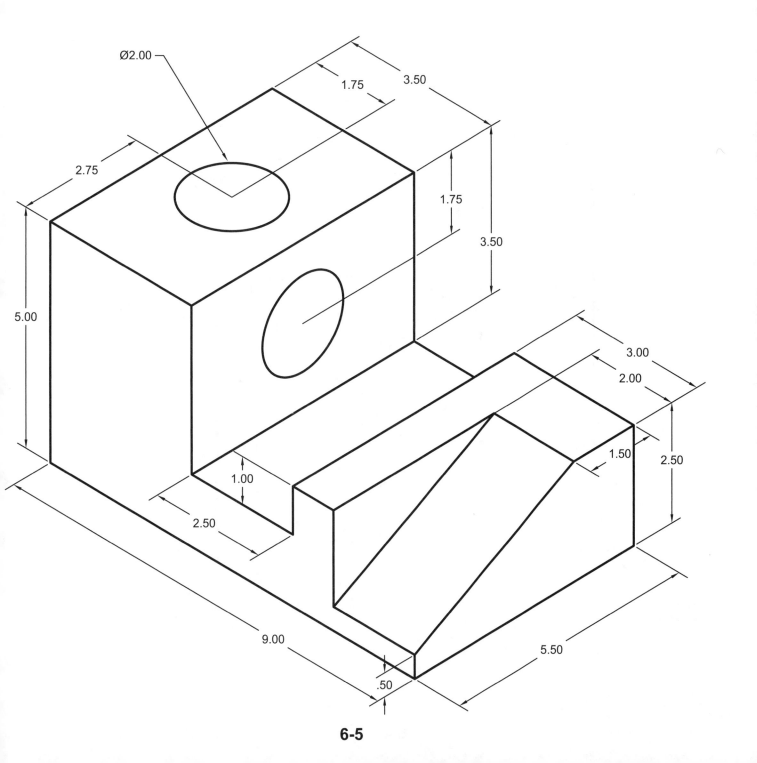

EXERCISE 6C

ABSTRACT HOUSE

1. Start a **New** file using **My Feet-Inches Setup.dwt**.
2. Select the **Model** tab
3. Set the **"Snap type and Style"** to isometric snap (Refer to page 6-2)
4. Change to isoplane with F5 when necessary.
5. Draw the Abstract House shown below. Do not dimension
6. Save as: **EX-6C**

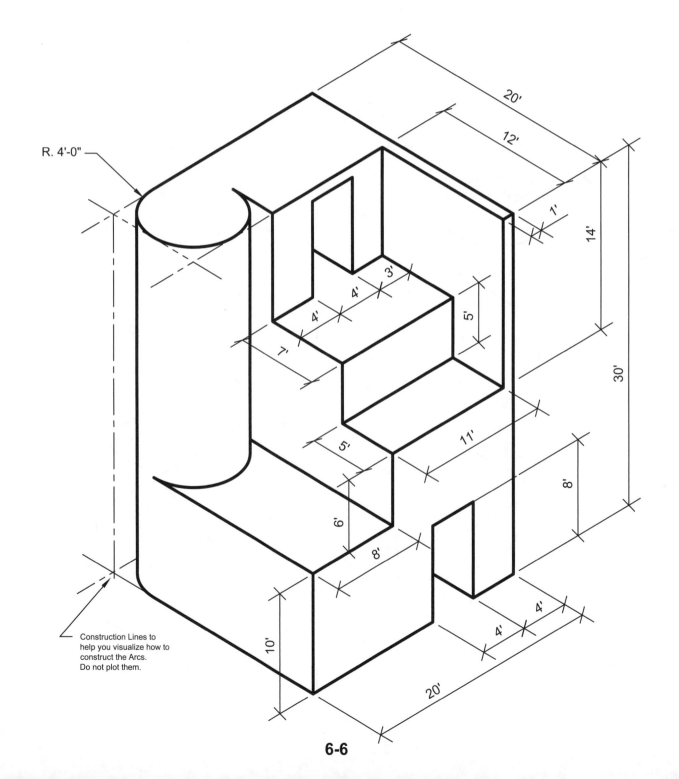

R. 4'-0"

Construction Lines to
help you visualize how to
construct the Arcs.
Do not plot them.

20'
12'
1'
14'
30'
3'
4'
4'
5'
7'
5'
11'
8'
6'
8'
10'
4'
4'
20'

LEARNING OBJECTIVES

After completing this lesson, you will be able to:

1. Copy, Cut and Paste
2. Dimension an Isometric object
3. Create Isometric Text

LESSON 7

Copy Clip and Cut

AutoCAD allows you to **Copy** or **Cut** objects from one drawing and **Paste** them into another drawing or document.

Copy Clip (copies the selected objects to clipboard, base point defaults to 0,0)

1. Select the command using one of the following:

    Ribbon = Home tab / Clip board panel /

    Short cut Menu (right click) = Copy

    Keyboard = Ctrl + C

2. Copy Clip: Select the objects that you want to copy.

3. The selected objects will be copied to the clipboard (Computers memory)

Copy with Base point (allows you to select a base point and copies the selected objects to clipboard)

1. Select the command using one of the following:

    Ribbon = None

    Short cut Menu (right click) = Copy with Base Point

    Keyboard = Ctrl + Shift + C

2. Copy with Base Point: **First** select the base point and **then** select the objects to copy

3. The selected objects will be copied to the clipboard (Computers memory)

Cut (copies the selected objects to the clipboard and deletes them from drawing)

1. Select the command using one of the following:

    Ribbon = Home tab / Clip board panel /

    Short cut Menu = Cut

    Keyboard = Ctrl + X

2. Select the objects you wish to Copy and Delete (at the same time)

3. The selected objects will be copied to the clipboard and then deleted from the drawing.

Paste

After you have used one of the copy commands, on the previous page, you may **paste** the previously selected objects into the current drawing, another drawing or a document created with another software such as MS Word or MS Excel.

1. Select the command using one of the following:

    Ribbon = Home tab / Clip board panel / **(See other options below)**

    Short cut Menu (right click) = Paste

    Keyboard = Ctrl + V

2. Place the objects in the location desired.

ADDITIONAL PASTE OPTIONS:

 PASTE as BLOCK or PASTEBLOCK
Copies objects into the same or different drawing as a BLOCK.

 PASTE as a Hyperlink

 PASTE to ORIGINAL COORDINATES or PASTEORIG
Pastes objects into the new drawing at the same coordinate position as the original drawing. You will not be prompted for an insertion point.

 PASTE SPECIAL or PASTESPEC
Use when pasting from other applications into AutoCAD.

How to Copy and Paste

To copy selected objects from one drawing and Paste into another drawing is very easy. There are just a few things new things you have to learn.

How to copy objects from one drawing to another drawing.

Step 1. Copy the objects to the clipboard.

1. Make sure that SDI is set to 0. (Refer to page 1-2)
2. Open 2 or more drawings.
3. Select **Vertical** from the Window panel. (Refer to page 1-3)
4. Click on the drawing from which you will copy objects to make the drawing **Active**.

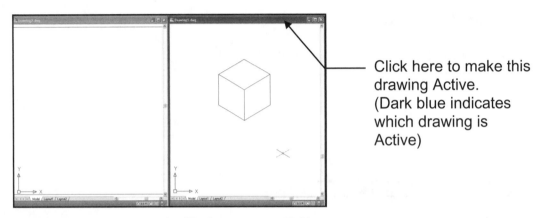

Click here to make this drawing Active.
(Dark blue indicates which drawing is Active)

5. Select the Copy Clip command. (Refer to page 7-2)
6. Select the objects to copy to the clip board

Step 2. Paste the Objects into another drawing.

1. Click on the drawing in which you will paste the previously copied objects to make the drawing **Active**.
2. Select the Paste command. (Refer to page 7-3)
3. Place the previously copied objects in the desired location.

Notice the drawing may appear a different size. It depends on the scale of the viewport or Zoom. But the actual size has not changed

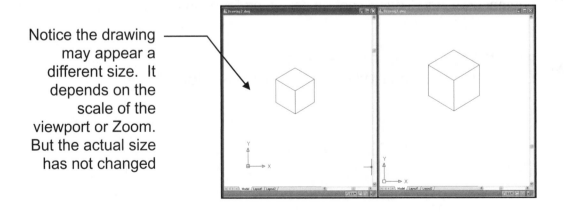

ISOMETRIC TEXT

AutoCAD doesn't really have an Isometric Text. But by using both **Rotation** and **Oblique angle**, we can make text appear to be laying on the surface of an Isometric object.

1. First create two new text styles. One with an <u>oblique angle</u> of **30** and the other with an <u>oblique angle</u> of **minus 30**.

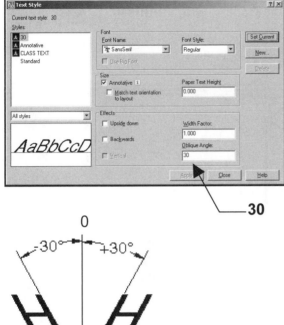

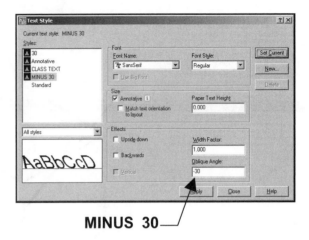

— 30   MINUS 30 —

2. Next select the appropriate text style and rotation as follows:

   a. Select the text style with an obliquing angle of 30 or minus 30.
   b. Select **SINGLE LINE TEXT**
   c. Place the START POINT or Justify.
   d. Type the Height.
   e. Type the Rotation angle.
   f. Type the text.

This is an example of what you can achieve. The text appears to be on the top and side surfaces.

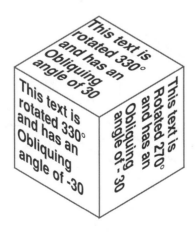

Dimensioning an Isometric Drawing

Dimensioning an isometric drawing in AutoCAD is a two-step process.
First you dimension it with the dimension command **ALIGNED**.
Then you adjust the angle of the extension line with the dimension command
OBLIQUE.

Step 1. Dimension the object using the dimension command **Aligned.**

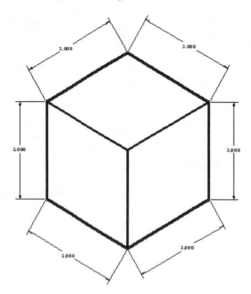

Step 2. Adjust the angle of the **extension lines** using the dimension command
Oblique. (Do each dimension individually)

1. Select dimension command **Oblique** ⊢⊣

2. Select the dimension you wish to adjust.
3. Enter the oblique angle for the extension line. (30, 150, 210 or 330)

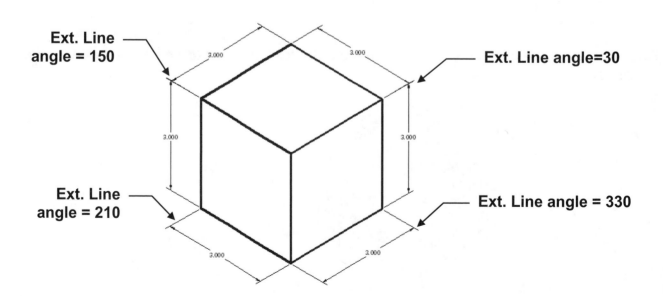

Ext. Line angle = 150

Ext. Line angle=30

Ext. Line angle = 210

Ext. Line angle = 330

EXERCISE 7A

OBLIQUE DIMENSIONING

1. Open drawing **EX-6B**
2. Dimension the isometric object shown below. (Refer to page 7-6)
3. Use "Aligned" and then "Oblique".
4. Change to isoplane "**Top**"
5. Use grips to move dimensions if necessary.
6. Save as: **EX-7A**

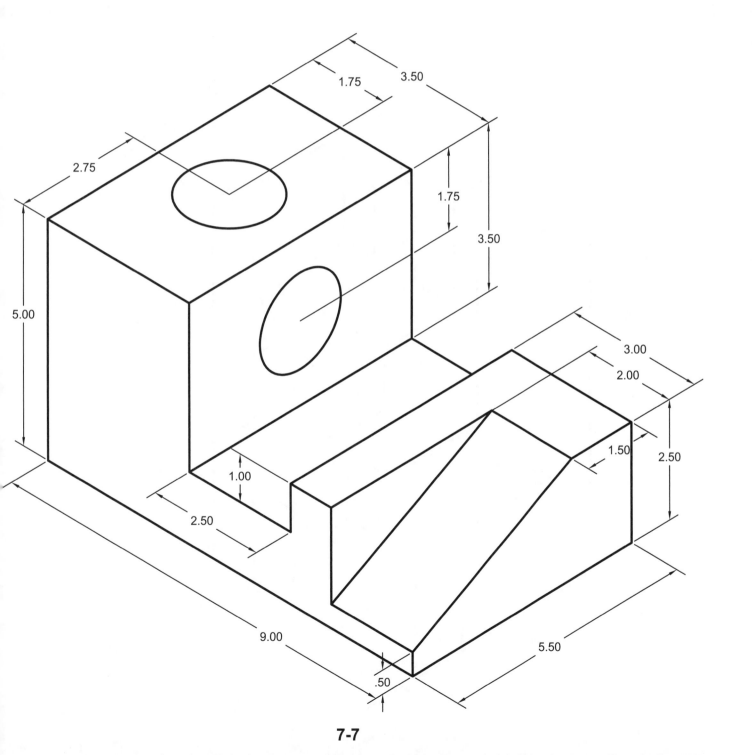

EXERCISE 7B

OBLIQUE DIMENSIONING

1. Open drawing **EX-6C**
2. Dimension the house.
3. Save as: **EX-7B**

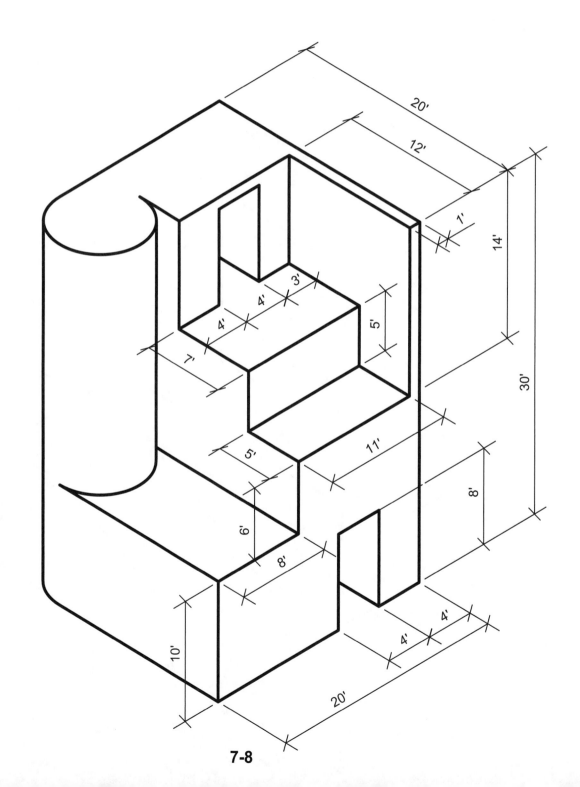

EXERCISE 7C

ISOMETRIC TEXT

1. Start a **New** file using **My Decimal Setup.dwt**
2. Select the model tab.
3. First create two new text styles. One with an oblique angle of 30 and the other with an oblique angle of minus 30.

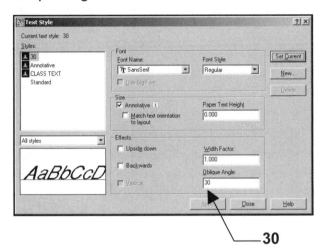

30

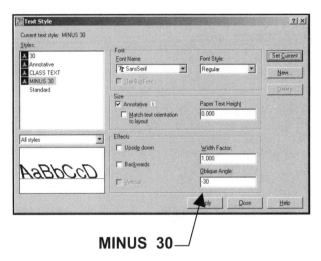

MINUS 30

4. Draw a 4 inch cube.

5. Next select the appropriate text style and rotation to create the text on the cube shown below.

   a. Select the text style with an oblique angle of 30 or minus 30.
   b. Select **SINGLE LINE TEXT**
   c. Place the START POINT (Lower left corner) approximately as shown.
   d. Type the Height = .25
   e. Type the Rotation angle.
   f. Type the text.

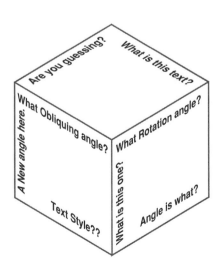

Notes:

LEARNING OBJECTIVES

After completing this lesson, you will be able to:

1. Review Creating a Block
2. Review inserting a Block
3. Assign and use Attributes

LESSON 8

BLOCKS

You learned the **Block** command in the Beginning Workbook. In this lesson you will learn how to add **Attributes** to a block and when to make your Blocks **Annotative.** I have included the instructions on **Creating a Block** as a review. If you feel that you thoroughly understand creating and inserting blocks, skip to page 8-6.

CREATING A BLOCK

1. First draw the objects that will be converted into a Block.

    For this example a circle and 2 lines are drawn.

2. Select the **BLOCK MAKE** command using one of the following:

    Ribbon = Insert tab / Block panel /

    Keyboard = B <enter>

3. Enter the New Block name in the **Name** box.

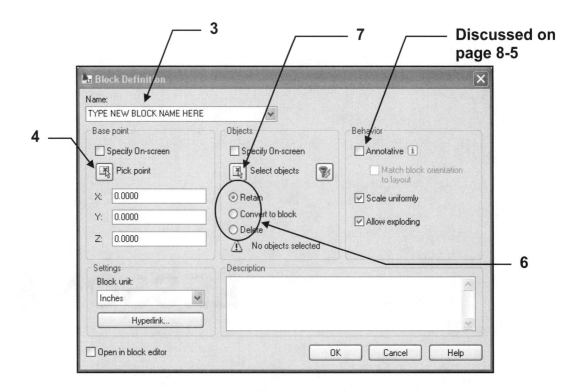

BLOCKS....continued

4. Select the **Pick Point** button. (Or you may type the X, Y and Z coordinates.)
   The Block Definition box will disappear and you will return temporarily to the drawing.

5. Select the location where you would like the insertion point for the Block.
   Later when you insert this block, the block will appear on the screen attached to the cursor at this insertion point. Usually this point is the CENTER, MIDPOINT or ENDPOINT of an object.

Notice the coordinates for the base point are now displayed. (Don't worry about this. Use Pick Point and you will be fine)

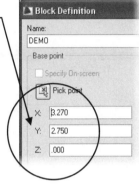

6. Select an option.

   It is important that you select one and understand the options below.

   Retain
   If this option is selected, the original objects will stay visible on the screen after the block has been created.

   Convert to block
   If this option is selected, the original objects will disappear after the block has been created, but will immediately reappear as a block. It happens so fast you won't even notice the original objects disappeared.

   Delete
   If this option is selected, the original objects will disappear from the screen after the block has been created. (This is the one I use most of the time)

7. Select the **Select Objects** button. (Note: Do not use the QuickSelect button.)

   The Block Definition box will disappear and you will return temporarily to the drawing.

8. Select the objects you want in the block, then press <enter>.

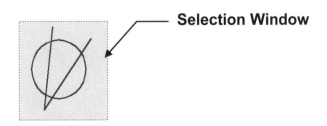

Selection Window

BLOCKS....continued

The Block Definition box will reappear and the objects you selected should be illustrated in the Preview Icon area.

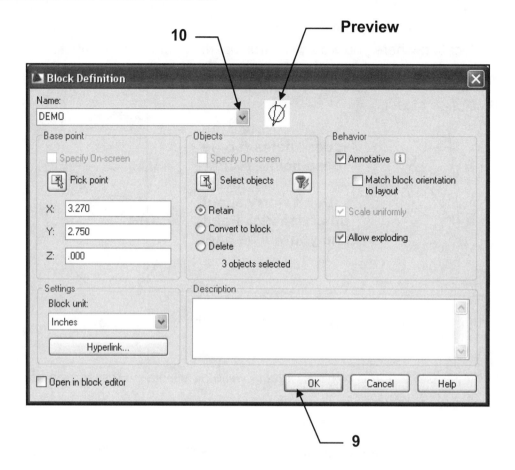

9. Select the **OK** button.
   The new block is now stored in the drawing's block definition table.

10. To verify the creation of this Block, select **Make Block** command again, select the Name (▼). A list of all the blocks, in this drawing, will appear.

ANNOTATIVE BLOCKS

If you want the appearance and size of the a Block to remain consistent, no matter what the scale of the viewport, you may make it <u>annotative</u>. Normally Blocks are not Annotative. For example, a Bathtub would not be annotative. You would want the Bathtub appearance to increase or decrease in size as you change the viewport scale. But the size of a Page or Section designator, as shown below, may be required to remain a consistent size no matter where it is displayed. To achieve this, you would make it Annotative.

Example:
1. Make a block such as the example below.

2. Insert the block into your drawing in modelspace.
3. Add an **annotative object scale** such as: 1:2 or 1:4 to the block.
     (Refer to the Beginning workbook lesson 28 to review how to add multiple annotative scales)

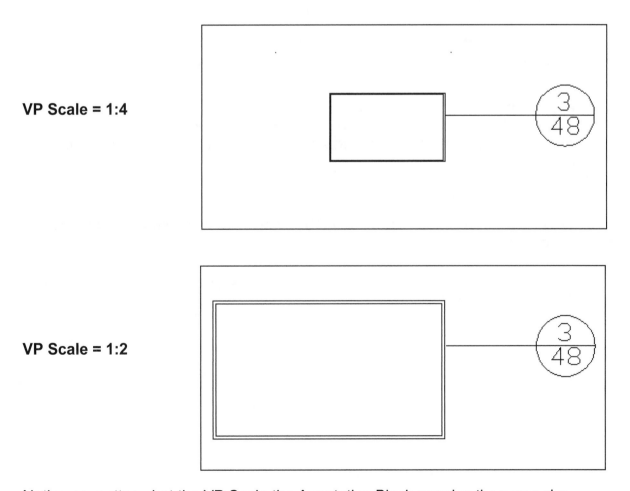

VP Scale = 1:4

VP Scale = 1:2

Notice no matter what the VP Scale the Annotative Block remains the same size.

REVIEW HOW TO INSERT A BLOCK

A **BLOCK** can be inserted at any location within the drawing. When inserting a Block you can **SCALE** or **ROTATE** it.

1. Select the **INSERT** command using one of the following:

   Ribbon = Insert tab / Block panel /

   Keyboard = Insert <enter>

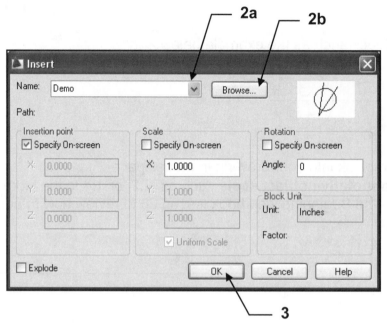

2. Select the **BLOCK** name.
   a. If the block is already in the drawing that is open on the screen, you may select the block from the drop down list shown above
   b. If you want to insert an entire drawing, select the Browse button to locate the drawing file.

3. Select the **OK** button.

   This returns you to the drawing and the selected block should be attached to the cursor.

4. Select the insertion location for the block by moving the cursor and pressing the left mouse button or typing coordinates.

ATTRIBUTES

The **ATTRIBUTE** command allows you to add text data to a block. You define the attribute and attach it to a block. Every time you insert the block, the attributes are also inserted.

For example, if you had a block in the shape of a tree, you could assign information (attributes) about this tree, such as name, size, cost, etc.   Each time you insert the block, AutoCAD will pause and prompt you for "What kind of tree is this?" You respond by entering the tree name. Then another prompt will appear, "What size is this tree? You respond by entering the size. Then another prompt will appear, "What is the cost"? You respond by entering the cost. The tree symbol will then appear in the drawing with the name, size and cost displayed.

Another example could be a title block. If you assign attributes to the text in the title block, when you insert the title block it will pause and prompt you for "What is the name of the drawing?" or "What is the drawing Number?". You respond by entering the information. The title block will then appear in the drawing with the information already filled in.

You will understand better after completing the following.

CREATING BLOCK ATTRIBUTES

1. Draw a rectangle 2 " X 1"

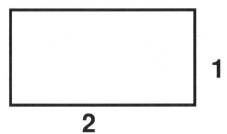

2. Select the **Define Attribute** command using one of the following:

    Ribbon = Insert tab / Attributes Panel / Define Attributes

    Keyboard = ATTRIBUTE

The following dialog box will appear.

**Definitions
on page 8-10**

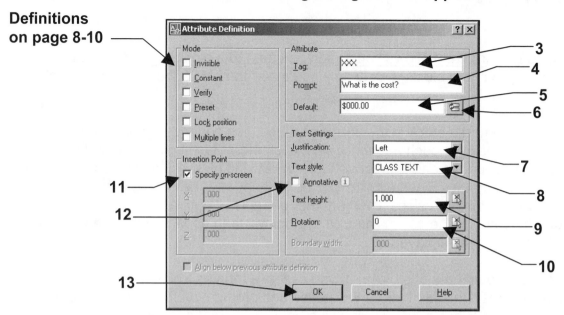

3. Enter the Attribute tag.
   A "Tag" is only a place saver. The "XXX's" will be replaced with your answer to the prompt when the block is inserted. No spaces allowed.

4. Enter the Attribute prompt.
   This is the prompt that will appear when AutoCAD pauses and prompts you for information. (You create the prompt) If you select Mode "constant" the prompt will not appear.

5. Enter a "default" value if necessary.
   The value will appear beside the prompt, in brackets, indicating what format the prompt is requesting. Example: What is the cost? <$0.00> The $0.00 inside the brackets is the "Value".

6. Insert Field Button

   Displays the Field dialog box. You can insert a field as all or part of the value for an attribute. Fields were discussed previously in Lesson 5.

7. Select the Justification for the text.
   This will be used to place the Attribute text.

8. Select a Text Style.
   The Text Style selected will be used when displaying the Attribute.

9. Enter the text Height.
   This will be the height of the Attribute text.

10. Enter the Rotation angle.
    This will be the rotation angle of the Attribute text. Such as: 90, 270 etc.

11. Insertion Point.
    You may uncheck the "Specify on Screen" box and enter the X, Y and Z coordinates if you know the specific coordinates or select the "Specify on Screen" box to place the insertion point manually.

12. Select the **OK** button.

13. The "**tag**" will be attached to the cursor.  It is waiting for you to place it somewhere on the drawing to establish the insertion point.

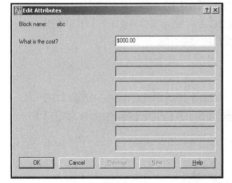

Snap to the corner

Pick Point

14. Select **Block Make** command to create a block. (Refer to page 8-2)
    Make a block of the rectangle and the attribute text.
    Name = abc
    Pick Point = upper right corner of rectangle
    Select Objects = Select rectangle and attribute text with a crossing window.
    Annotative = Do not make it annotative

15. Select **OK** button.

16. Select **INSERT** command (Refer to page 8-6)
    a. Select block name **abc** and then **OK** button.
    b. Place the block somewhere on the screen and press the left mouse button.
       The "Enter Attributes" prompt will appear on the command line.
    c. Type: **$300.00 <enter>**

If you prefer the EDIT ATTRIBUTES DIALOG BOX

Edit Attributes

Block name: abc

What is the cost? $000.00

OK   Cancel   Previous   Next   Help

Set the "ATTDIA" variable to 0 or 1 as follows:
Type: Attdia <enter>
Enter 0 or 1 <enter>

If the variable **ATTDIA** is set to **0**, the prompt will appear on the command line.

If the variable **ATTDIA** is set to **1**, the dialog box shown here will appear.

It is strictly personal preference. It will not affect the input.
(Personally, I prefer the "dialog box")

---

17. Select **INSERT** command **again.**
    a. Select block name **abc** and then **OK** button.
    b. Place the block somewhere on the screen and press the left mouse button.
    c. Type: **$450.00 <enter>**

18. Select **INSERT** command **again.**
    a. Select block name **abc** and then **OK** button.
    b. Place the block somewhere on the screen and press the left mouse button.
    c. Type: **$75.00 <enter>**

You should now have 3 rectangles that look like the ones shown below. You inserted the same block but the text is different in each. Think how this could be useful in other applications.

$300.00 $450.00 $75.00

ATTRIBUTE MODES

Invisible
If this box is checked, the attribute will be invisible. You can make it visible later by typing the ATTDISP command and selecting ON.

Constant
The attribute stays constant. It never changes and you will not be prompted for the value when the block is inserted.

Verify
After you are prompted and type the input, you will be prompted to verify that input. This is a way to double check your input before it is entered on the screen. This option does not work when ATTDIA is set to 1 and the dialog box appears.

Preset
You will not be prompted for input. When the block is inserted it will appear with the default value. You can edit it later with **ATTEDIT**. (See Editing Attributes page 9-2.)

Lock Position
Locks the location of the attribute within the block reference. When unlocked, the attribute can be moved relative to the rest of the block using grip editing, and multiline attributes can be resized.

Multiple Lines
Specifies that the attribute value can contain multiple lines of text. When this option is selected, you can specify a boundary width for the attribute.

EXERCISE 8A

Assigning Attributes to a Block

The following exercise will instruct you to draw a Box, Assign Attributes, create a block of the Box including attributes, then insert the new block and answer the prompts when they appear on the screen.

1. Open **My Decimal Setup** and select the **MODEL** tab.

2. Draw the isometric Box shown on the next page using the following dims.
   L = 3.75   W = 2.25   H = 3.00

3. Assign Attributes for **Length**
   a. Tag = LENGTH
   b. Prompt = What is the Length of the Box?
   c. Default = inches
   d. Justification = Middle
   e. Text Style = CLASS TEXT
   f. Height = .250
   g. Rotation = 30
   h. Annotation = off
   i. Select OK and place as shown on 8-12.

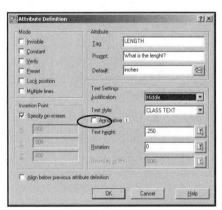

4. Assign Attributes for **Width**.
   a. Tag = WIDTH
   b. Prompt = What is Width of the Box?
   c. Default = inches
   d. Justification = Middle
   e. Text Style = CLASS TEXT
   f. Height = .250
   g. Rotation = 330
   h. Annotation = off
   i. Select OK and place as shown on 8-12.

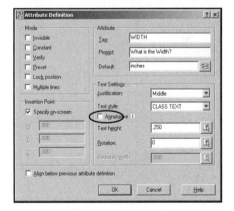

5. Assign Attributes for **Height**.
   a. Tag = HEIGHT
   b. Prompt = What is Height of the Box?
   c. Default = inches
   d. Justification = Middle
   e. Text Style = CLASS TEXT
   f. Height = .250
   g. Rotation = 90
   h. Annotation = off
   i. Select OK and place as shown on 8-12.

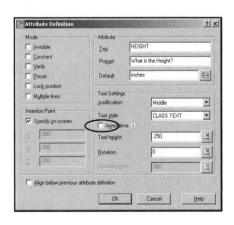

6. Assign Attributes for **Manufacturer**.
   a. Tag = MANUFACTURER
   b. Prompt = Who is the MFR?
   c. Default = (leave blank)
   d. Justification = Left
   e. Text Style = CLASS TEXT
   f. Height = .250
   g. Rotation = 30
   h. Annotation = No
   i. Select OK and place as shown below.

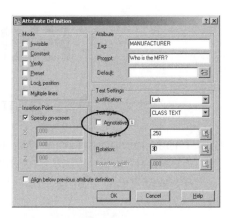

7. Assign Attributes for **Cost**.
   a. Tag = $$$$
   b. Prompt = What is the cost?
   c. Default = $0.00
   d. Annotation = Off
   e. Select "Align below previous attribute definition" box.
   f. Select OK. But you do not have to place the attribute tag. It automatically "Aligned below previous attribute tag, MANUFACTURER".

Notice grayed out

Your drawing should look approximately like the example below.

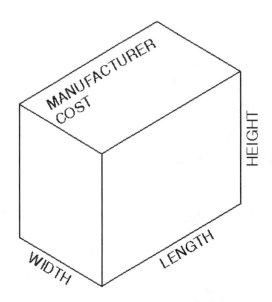

8. When you have finished assigning all of the attributes, create a **BLOCK**. Select the Box and the Attribute text. (Refer to page 8-2 for instructions if necessary)

9. Now **Insert** the new block anywhere on the screen using
   a. Select **INSERT**
   b. Select the block from drop down list.
   c. Select OK

b

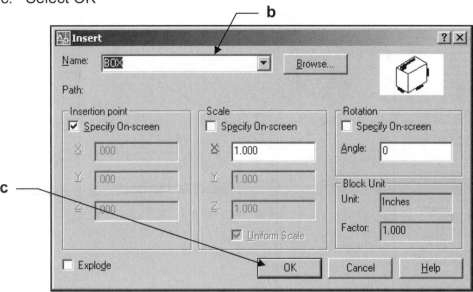

10. Type the answers in the box beside the attribute prompts as shown below.

Note: If the dialog box shown below does not appear, at the command line type ATTDIA <enter> and 1 <enter>.

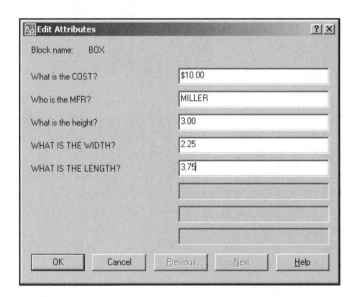

Notice that the prompts are not in any specific order. You can control the initial order of the prompts, when creating the block, by selecting the attribute text one by one with the cursor instead of using a window to select all objects at once.
In Lesson 9 you will learn how to rearrange the order after the block has been created.

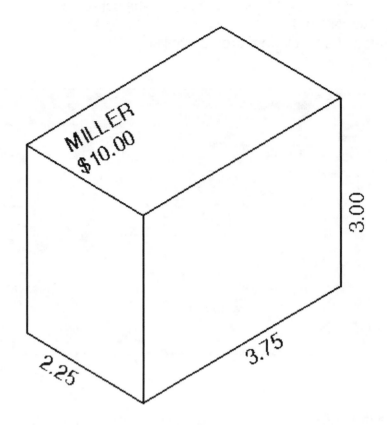

Does your box look like the example shown above?

11. Save as **EX-8A.**

EXERCISE 8B
CREATE A FLOOR PLAN WITH BLOCKS AND ATTRIBUTES

A. Select **My Feet-Inches Setup .dwt**

B. Select the **A Size** tab.

C. Unlock the Viewport.

D. Set the Viewport scale to **3/16" = 1'**

E. Lock the Viewport

G. <u>Draw the Floor Plan</u> shown on the next page.
   1. The exterior walls are 6" thick
   2. The interior walls are 4" thick
   3. Use Layer Walls
   4. Leave spaces for the Windows and Doors OR use trim or break.

H. <u>DOORS</u>
   1. Use Layer DOORS
   2. Door Size = 30" (4" space behind door)

I. <u>FURNITURE</u>
   1. Use Layer FURNITURE
   2. Place approximately as shown.
   3. Sizes:
      Credenza's = 18" x 5'
      Lamp Table = 2' Sq.
      Copier = 2' x 3'

J. <u>ELECTRICAL</u>
   1. Use Layer ELECTRICAL for switches and fixtures.
   2. Use Layer **WIRING** for the wire from switches to fixtures.
   3. Sizes:
Wall Outlet = 8" dia.   Overhead Lights = 16" dia.   Switches = Text ("S" ht = 1/8")

K. <u>AREA TITLES</u> (Office, Lobby and Reception) in PAPERSPACE.
   1. Use <u>Layer</u> = Text Heavy   <u>Text Style</u> = Arch Text   <u>Height</u> = 3/16" in Modelspace.

L. <u>DIMENSION</u> in Modelspace.
   a. Use Dimension Style "Arch Dim"
   b. Annotation scale = same as Viewport scale

M. <u>Save as</u>: Ex-8B

EXERCISE 8B continued

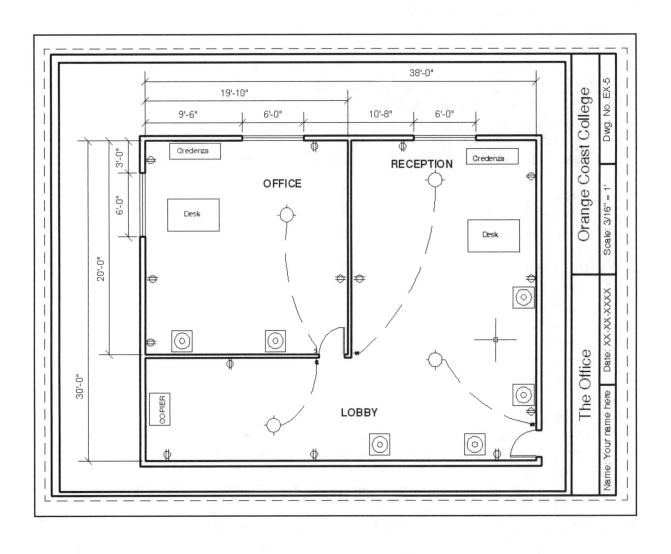

EXERCISE 8C

Assigning multiple Attributes to multiple Blocks

The following exercise will help you learn how to assign multiple attributes to 3 different objects. This drawing will be used in lesson 9 to extract the information and place it in a spreadsheet such as Excel.

A. Open **EX-8B**

B. Draw the additional furniture shown below. (Chair, sofa and file cabinet)

    Chair = 2' square   Sofa = 6' X 3'   File cabinet = 24" X 15"

C. Assign Attributes to each and create blocks using Step 1 and Step 2 on the following pages.

C. START WITH THE SOFA

This is what the sofa should look like after you have completed the Attribute definitions shown below.

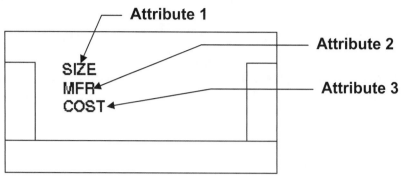

STEP 1.

1. Select **Define Attributes**
2. Fill in the boxes for Attribute 1 as shown below.
3. Select the OK button and place the Tag approximately as shown above.
4. Repeat 1, 2 and 3 for Attributes 2 and 3.
5. Then go on to **Step 2** in the lower right corner of this sheet.

ATTRIBUTE 1

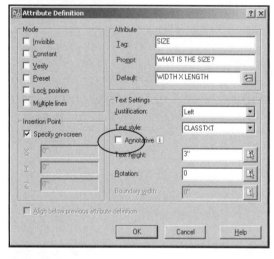

ATTRIBUTE 2

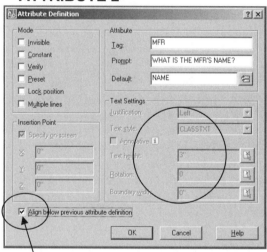

ATTRIBUTE 3

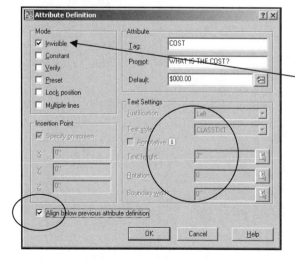

This option will align the new attribute directly below the previous and duplicate the Text options.

Note: the tag will not be invisible until you insert the block.

STEP 2

5. Now **create a block**.
a. Select **Create**
b. Name = Sofa
c. Select objects = select the sofa and the attribute text.
d. Pick Point = Any corner of the sofa.

D. NOW DO THE CHAIR

This is what the chair should look like after you have completed the Attribute definitions shown below.

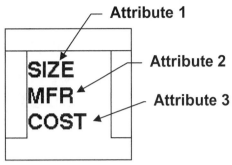

STEP 1

1. Select **Draw / Block / Define Attributes**
2. Fill in the boxes for Attribute 1.
3. Select the **OK** button and place the Tag (Size) approximately as shown.
4. Repeat steps 2 and 3 for Attributes 2 and 3.
5. Then go on to **Step 2** in the lower right corner of this page.

ATTRIBUTE 1

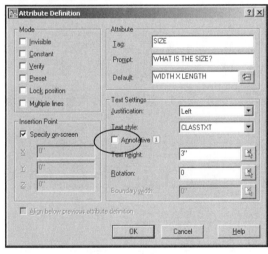

ATTRIBUTE 2

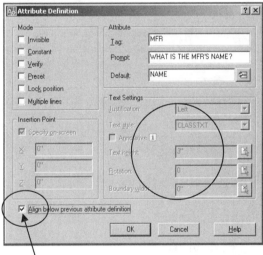

This option will align the new attribute directly below the previous and duplicate the Text options.

Note: the tag will not be invisible until you insert the block.

ATTRIBUTE 3

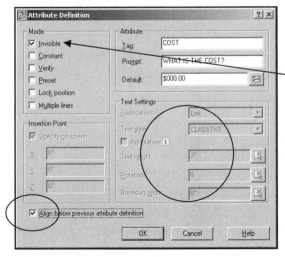

STEP 2

5. Now **create a block**.
a. Select **Create**
b. Name = Chair
c. Select objects = select the sofa and the attribute text.
d. Pick Point = Any corner of the sofa.

E. __NOW DO THE FILE CABINET__

This is what the file cabinet should look like after you have completed the Attribute definitions shown below.

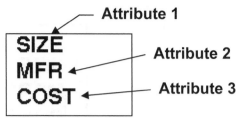

SIZE ← Attribute 1

MFR ← Attribute 2

COST ← Attribute 3

STEP 1
1. Select **Draw / Block / Define Attributes**
2. Fill in the boxes for Attribute 1.
3. Select the **OK** button and place the Tag (Size) approximately as shown.
4. Repeat steps 2 and 3 for Attributes 2 and 3.
5. Then go on to **Step 2** in the lower right corner of this page.

ATTRIBUTE 1

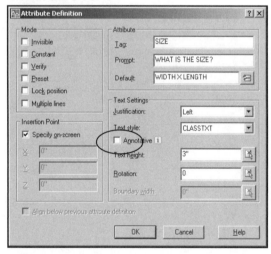

ATTRIBUTE 2

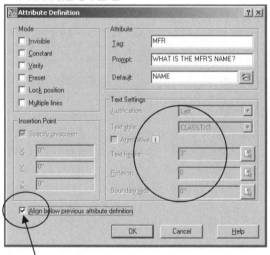

ATTRIBUTE 3

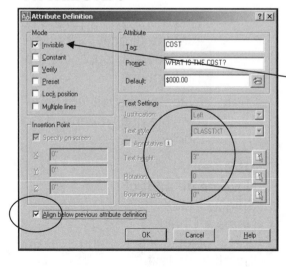

This option will align the new attribute directly below the previous and duplicate the Text options.

Note: the tag will not be invisible until you insert the block.

STEP 2
5. Now **create a block**.
a. Select **Create**
b. Name = File Cabinet
c. Select objects = select the sofa and the attribute text.
d. Pick Point = Any corner of the sofa.

Save this drawing as EX-8C now so you don't lose anything.

8-20

F. INSERT the **Sofa block** as shown on page 8-22
    1. Select **INSERT**
    2. Select the **SOFA** block from the drop down list.
    3. Select **OK**
    4. Answer the Attribute prompts:

        What is the SIZE?   3' X 6'
        Who is the Manufacturer? Sears
        What is the Cost? $500.00

    5. Select **OK**.

G. INSERT the **Chair block** as shown on page 8-22

    1. Select **INSERT**
    2. Select the **CHAIR** block from the drop down list.
    3. Select **OK**
    4. Answer the Attribute prompts:

        What is the SIZE?   2' X 2'
        Who is the Manufacturer? LAZYBOY
        What is the Cost? $200.00

    5. Select OK.

H. INSERT the **File cabinet block** as shown on page 8-22

    1. Select **INSERT / BLOCK**
    2. Select the **File Cabinet** block from the drop down list.
    3. Select **OK**
    4. Answer the Attribute prompts:

        What is the SIZE?   15" X 24"
        Who is the Manufacturer? HON
        What is the Cost? $40.00

    5. Select **OK**.

Your drawing should look approximately like the example shown below.

I. Save this drawing as: **EX-8C**

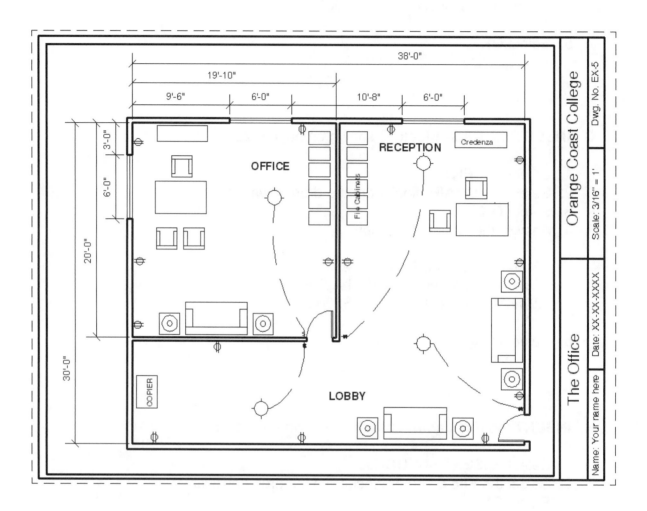

LEARNING OBJECTIVES

After completing this lesson, you will be able to:

1. Edit Attributes
2. Extract Attributes
3. Extract data to an AutoCAD table.
4. Extract data to an External file.

LESSON 9

EDITING ATTRIBUTES

After a block with attributes has been inserted into a drawing you may want to edit it. AutoCAD has many ways to edit these attributes.

> *Sorry LT users, you can __only__ use the ATTEDIT and ATTDISP commands shown on page 9-4 to edit the attribute text.*

1. Select the **BLOCK ATTRIBUTE MANAGER** command using one of the following:

   Keyboard = BATTMAN

2. Select the block that you want to edit.

   The following dialog box will appear.

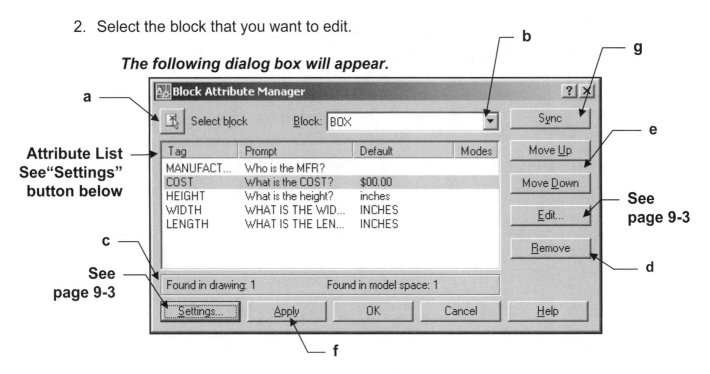

a. **Select Block** – Allows you to select another block – takes you back to the drawing so you can select another block to edit.

b. **Block down arrow** – Allows you to select another block –from a list.

c. **Found** – lists how many of the selected block were found in the drawing and how many are in the space you are currently in. (You must select the model "tab" or Layout "tab". It will not find model space attributes if you are not in model space.)

d. **Remove** - Allows you to remove an Attribute from a block. (See "Apply and Sync" below)

e. **Move Up and Down** - Allows you to put the prompts in the order you prefer.

f. **Apply** – After you have made all the changes, select the Apply button to update the attributes. (The Apply button will be gray if you have not made a change.)

g. **Sync** – If you explode a block, make a change and redefine the block, Sync allows you to update all the previous blocks with the same name.

SETTINGS

The Settings dialog box controls which attributes are displayed. On the previous page TAG, PROMPT, DEFAULT and MODE are displayed. TAG values are always displayed.

a. Emphasize duplicate tags
If this option is ON, any duplicate tags will display RED.

b. Apply changes to existing refs
If this option is ON, the changes will affect all the blocks that are in the drawing and all the blocks inserted in the future.

Note, very important: If you only want the changes to affect **future** blocks, do not select "Apply changes to existing ref".
If you want all blocks changed,
select the SYNC button. (see page 9-2)

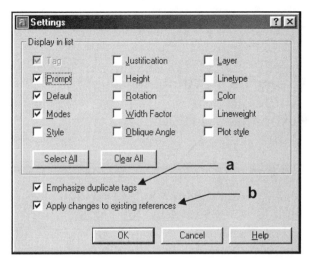

EDIT ATTRIBUTE

The Edit Attribute dialog box has 3 additional tabs, ATTRIBUTE, TEXT OPTIONS and PROPERTIES.

NOTE: If you select the **Auto preview changes** you can view the changes as you make them.

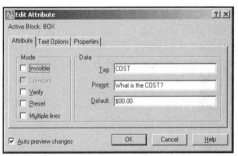

ATTRIBUTE tab
Allows you to change the MODE, TAG, PROMPT and DEFAULT.

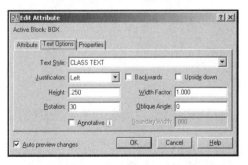

TEXT OPTIONS tab
Allows you to make changes to the Text.

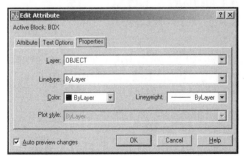

PROPERTIES tab
Allows you to make change to the properties.

WHAT IF YOU JUST WANT TO EDIT <u>ONE ATTRIBUTE</u> IN <u>ONE BLOCK</u>?

The following two commands allow you to edit the attributes in only one Block at a time and the changes will not affect any other blocks.

Select one of the commands below and then select the block that you wish to change.

To change the Attribute value:

Keyboard = ATTEDIT

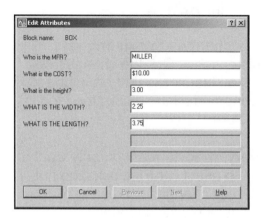

This command allows you to edit the attribute values for <u>only one block.</u>

This command will not affect other blocks.

ATTENTION LT USERS: You can't use EATTEDIT command.

To change the Attribute structure:

Keyboard = EATTEDIT
or double click on the block

This command allows you to edit the Attributes, Text Options, Properties and the Values of <u>only one block</u>.

This command will not affect other blocks.

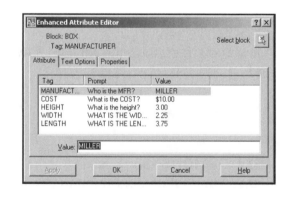

To change the visability of Attributes:

TYPE = ATTDISP
PULLDOWN = NONE
TOOLBAR = NONE

Normal = Retains the current visibility of each attribute. Visible attributes are displayed. Invisible attributes are not displayed.

On = Makes all attributes visible

Off = Makes all attributes invisible

WHAT IF YOU WANT TO <u>EDIT THE OBJECTS IN A BLOCK?</u>

If you would like to add, delete or change objects within an existing block, you may do it easily with the command **Refedit**.
Refedit allows you to make changes to a block, saves the changes and updates all other previously inserted blocks within the drawing automatically.
The changes only affect the current drawing.

1. Select the Refedit command using one of the following:

    Keyboard = Refedit

2. Select the Block you wish to edit. (Click on it, can't use a window)

The Reference Edit dialog box should appear.

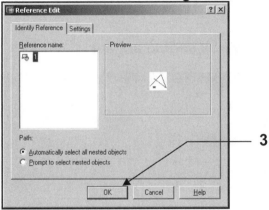

3

3. Select the OK button.

Note: The selected block will remain bold but all other blocks of the same name will fade to gray. This is to emphasis the selected block.

4. Make the changes to the selected block.

5. Select the **Edit Reference ▼**

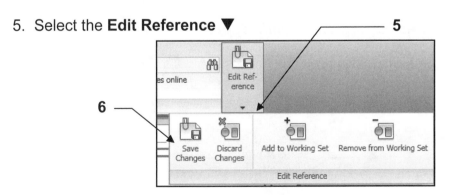

6. Select **Save Changes**

7. A warning will appear. Select the **OK** button.

The block has been redefined and all of the existing blocks have been updated to reflect the changes you made. (These changes affect the current drawing only)

EXTRACT DATA FROM BLOCK ATTRIBUTES

Sorry LT Users, you can't use this command.

Your blocks can now contain attribute information (data) such as size, manufacturer, cost or maybe even a bill of materials. The next step is to learn how to **extract** that **data** and save it to a table or to an external file such as a Microsoft Excel spread sheet.

This is a very simple process using the **Attribute Extraction Wizard**.

1. Select the **ATTRIBUTE EXTRACTION WIZARD**:

    Keyboard = DX

The "Data Extraction – Begin (Page 1 of 8)" dialog box should appear.

2. Select one of the following:

    Create a new data extraction:
    You may select "Use previous extraction as a template" if you have already created a template.

    Edit an existing data extraction

3. Select **Next >**

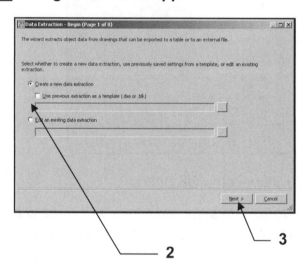

The "Save Data Extraction As" dialog box should appear.

4. Select the saving location and enter the new file name. Notice the extension will be .dxe

5. Select the **Save** button.

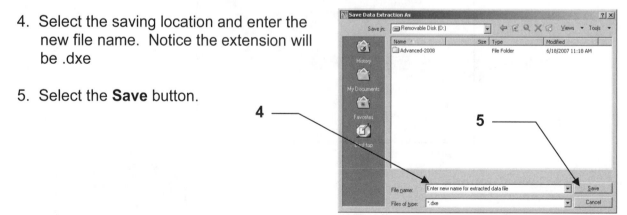

The Define Data Source (Page 2 of 8) dialog box should appear.

6. Select to extract data from the current drawing or selected objects only. The panel below gives the path of the current drawing.

7. Select **Next**

6 ——

7 ——

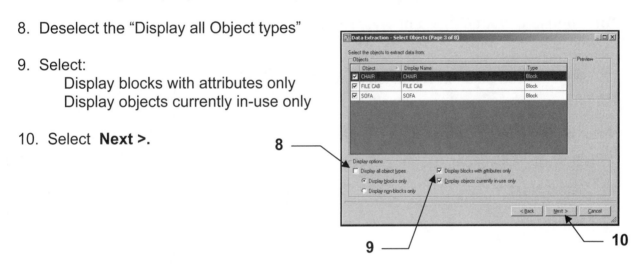

The Select Objects (Page 3 of 8) dialog box should appear.

8. Deselect the "Display all Object types"

9. Select:
   Display blocks with attributes only
   Display objects currently in-use only

10. Select **Next >.**

8 ——

9 ——

10 ——

The Select Properties (Page 4 of 8) dialog box should appear.

11. Deselect all categories except "Attribute"

12. Select **Next>.**

11 ——

12 ——

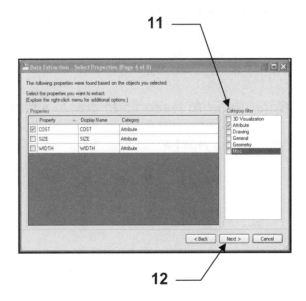

The Refine Data (Page 5 of 8) dialog box should appear

13. The display is controlled by the 3 options in the lower left corner.

14. Select **Next >**.

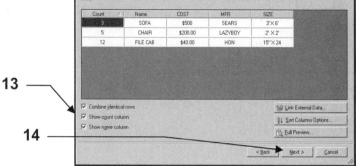

The Choose Output (Page 6 of 8) dialog box should appear

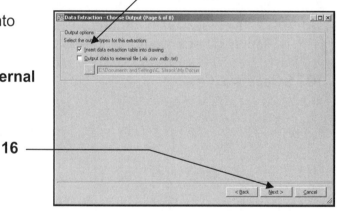

15. Select "Insert data extraction table into drawing".

Note: If you select "Output data to external file" refer to page 9-10

16. Select **Next >**

The Table Style (Page 7 of 8) dialog box should appear

17. Enter a "Title" for the table.

Note: Refer to lesson 5 to create a table.

18. Select **Next >**

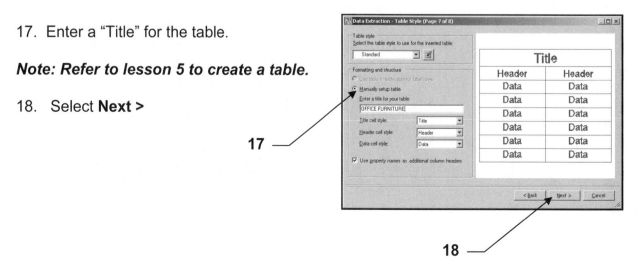

The Finish (Page 8 of 8) dialog box should appear

19. Select Finish

Read

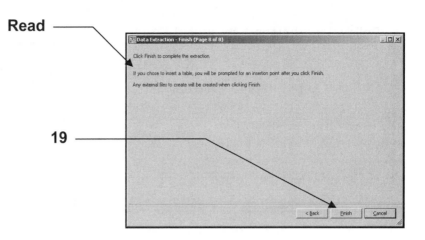

19

20. The Extracted Attribute data table should be attached to the cursor.
    You select the insertion location.
    You should insert the table into Paper space.
    The table will be inserted on the current layer.

OFFICE FURNITURE				
Count	Name	COST	MFR	SIZE
3	SOFA	$500	SEARS	3' X 6'
5	CHAIR	$200.00	LAZYBOY	2' X 2'
12	FILE CAB	$40.00	HON	15" X 24

To extract the attribute data to an External Spread sheet such as Microsoft Excel, refer to the next page.

EXTRACT DATA FROM BLOCK ATTRIBUTES TO AN EXTERNAL FILE.

2

3

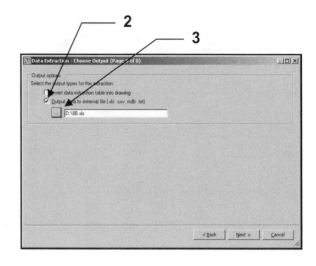

1. Follow instructions 1 through 14 on the previous pages.

2. Select "Output data to external file".

3. Select the […] button.

4. Locate where you wish to save the file.

5. Change **File of Type** to:

   .xls for Microsoft Excel

   .txt for Word Processing Program or Notepad.

6. Enter a name for the file.

7. Select **Save** button.

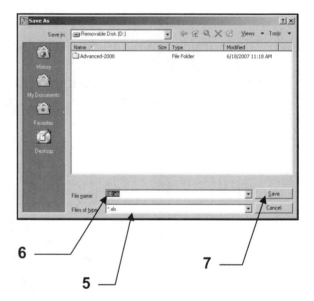

6

5

7

8. Select **Next >**

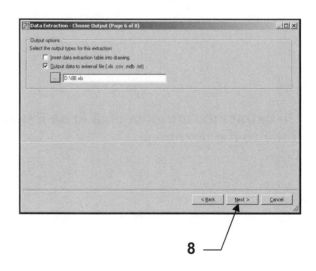

8

9. Select **Finish**.

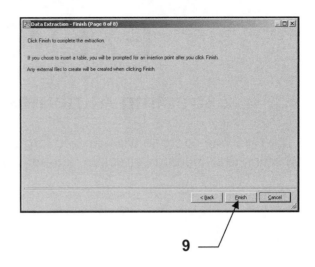

9

10. Open the External Program software that you selected in #5 on the previous page.

11. Select **File / Open**.

12. Locate the file.

Example of Microsoft Excel

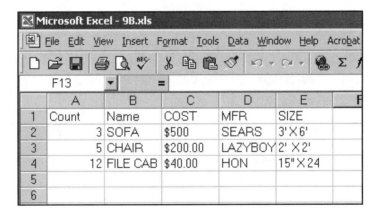

EXERCISE 9A

Extracting Attributes to an AutoCAD table.

The following exercise will take you through extracting attributes. You will open an existing drawing and extract the attribute data to an AutoCAD table.

1. Open **Ex-8B**.

2. Select the **A Size** layout tab.

3. Make sure you are in **Paper space**.

4. Follow the instructions on pages 9-6 through 9-9 to create the Table shown below and insert it into the drawing in paper space.

OFFICE FURNITURE				
Count	Name	COST	MFR	SIZE
3	SOFA	$500	SEARS	3' X 6'
5	CHAIR	$200.00	LAZYBOY	2' X 2'
12	FILE CAB	$40.00	HON	15" X 24

EXERCISE 9B

Extracting Attributes to an External File.

The following exercise will take you through extracting attributes. You will open an existing drawing and extract the attribute data to an external program.

1. Open **Ex-8B**.

2. Select the **A Size** layout tab.

3. Follow the instructions on page 9-10 through 9-11 to create the external data file shown below.

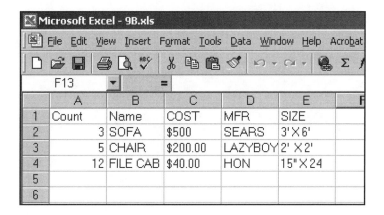

NOTES:

LEARNING OBJECTIVES

After completing this lesson, you will be able to:

1. Navigate in the DesignCenter Palette
2. Open a drawing from the DesignCenter Palette
3. Insert a block from the DesignCenter Palette
4. Drag and drop hatch patterns.
5. Drag and drop Symbols from the Internet

LESSON 10

DesignCenter

The AutoCAD **DesignCenter** allows you to find, preview and drag and drop Blocks, Dimstyles, Textstyles, Layers, Layouts and more, from the DesignCenter to an open drawing. You can actually get into a previously saved drawing file and copy any of the items listed above into an open drawing. This is a wonderful tool and easy to use.

Opening the DesignCenter palette

To open the **DesignCenter** palette select one of the following:

> **Ribbon** = **Insert tab / Content Panel /**
>
> **Keyboard = DC**

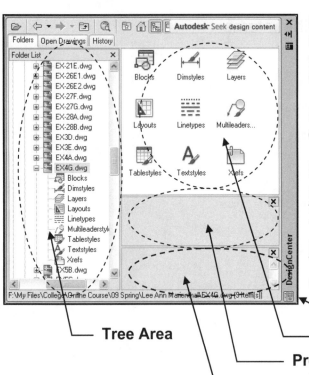

How to:

Resizing – You can change the width and height. Rest the cursor on an edge until the pointer changes to a double ended arrow. Click and drag to desired size.

Dock – Click the title bar, then drag it to either side of the drawing window.

Hide – You can hide the DesignCenter palette using the Auto-hide option. Click on the "properties" button and select "Auto-hide". When Auto-hide is ON, the palette is hidden, only the title bar is visible. The palette reappears when you place the cursor on the title bar.

Hide tool

Tree Area

Content Area

Preview Area

Description Area

VIEWING TABS
At the top of the palette there are 3 tabs that allow you to change the view.
They are: Folders, Open Drawings, and History.

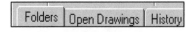

Folders tab – Displays the Directories and files similar to Windows Explorer. You can navigate and locate content anywhere on your system.

Open Drawings tab – Displays all open drawings. Allows you to select content from an open drawing and insert it into another open drawing. (Note: the target drawing must be the "active" drawing)

History tab – Displays the last 20 file locations accessed with DesignCenter. Allows you to double click on the path to load it into the "Content" area.

BUTTONS

At the top of the DesignCenter palette, there is a row of buttons. (Descriptions below) To select a button, just click once on it.

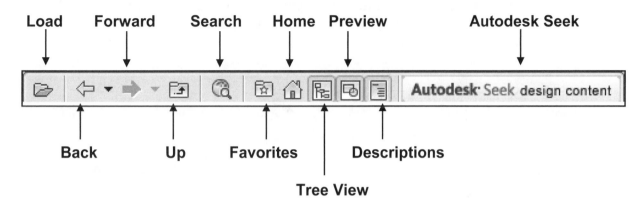

Load – This button displays the "Load" dialog box. (It is identical to the "Select File" dialog box.) Locate the drawing content that you want loaded into the "Content" area. You may also locate the drawing content using the Tree view and Folders tab.

Back and Forward – Allows you to cycle through previously selected file content.

Up – Moves up one folder from the current folder.

Search – Allows you to search for drawings by specifying various criteria.

Favorites – Displays the content of the Favorites folder. Content can be added to this folder. Right click over an item in the Tree View or Content areas then select "Add to Favorites" from the menu that appears. (You may add a drive letter, folder, drawing, layers, blocks etc.)

Home – Takes you to the DesignCenter folder by default. You can change this. Right click on an item in the Tree view area and select "Set as Home" from the menu. (You many select a drive letter, folder, drawing etc.)

Tree View – Toggles the Tree View On and OFF. Only works when the "Folders" or "Open Drawings" tab is current.

Preview – Toggles the "Preview" area On and OFF.

Description – Toggles the "Description" area On and OFF. (A description must have been given at the time the block was created.)

Autodesk Seek – Opens a web browser and displays the Autodesk Seek home page. Product design information available on Autodesk Seek depends on what content providers, both corporate partners and individual contributors, publish to Autodesk Seek. Such content could include 3D models, 2D drawings, specifications, brochures, or descriptions of products or components.

How to Insert a Block using DesignCenter.

There are two methods of inserting a block using DesignCenter.

Method 1: Drag and Drop
   a. Select "Blocks" from a specified drawing within the "Tree" area.
   b. Locate the Block desired in the Content area
   c. Drag and drop it into the drawing area of an open drawing.
      You will not be prompted for the insertion point.

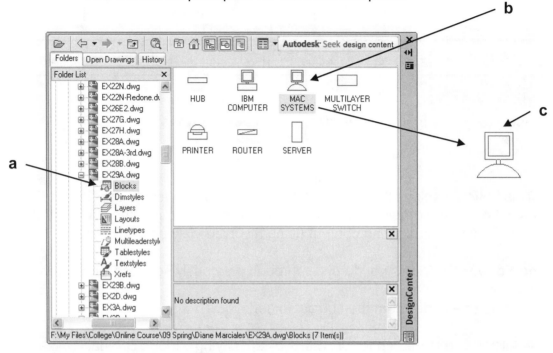

Method 2: Specify the Coordinates, Scale or Rotation
   Double click on the block in the Contents area. The Insert dialog box will appear. Specify coordinates for insertion point, scale factor and rotation angle.

To Close the DesignCenter palette
The DesignCenter palette will remain open until you close it.
(If you have "Autohide" ON, only the title bar will remain visible.)
To close the DesignCenter palette, select the **"X"** in the upper right or left corner of the Palette.

Drag and drop Layouts, Layers, Text Styles etc.

You can drag and drop just about anything from an existing file, listed in the Tree view "Folder List", to an open drawing file. This is a significant time saver. Just locate the source file in the tree area and drag and drop into an open drawing.

Some examples are shown below.

BLOCKS

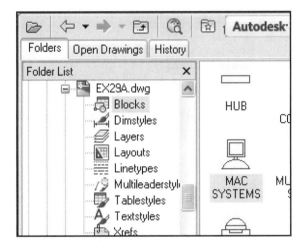

DIMENSION STYLES

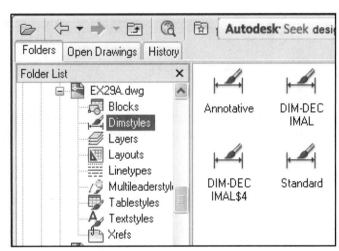

LAYERS

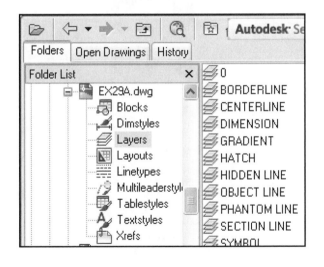

LAYOUTS

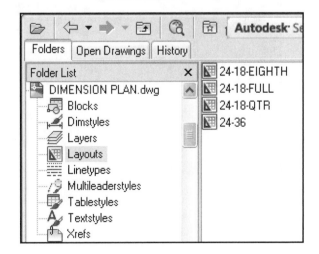

Autodesk Seek

Autodesk Seek connects you to the Internet and gives you access to thousands of symbols and manufacturer's product information.

Note: You must have an Internet connection to use this feature.

Just click on "Autodesk Seek" and AutoCAD automatically opens your Internet connection to the correct Internet address.

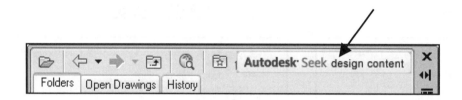

Experiment with Autodesk Seek. You may download varies drawings, symbols and tools from the architecture, engineering and construction markets.

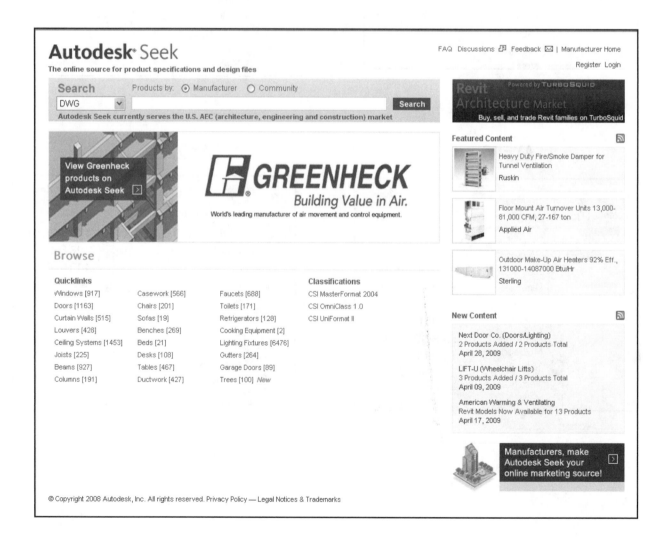

EXERCISE 10A

Inserting Blocks from the DesignCenter.

Step 1.

1. Start a NEW file using **My Feet-Inches Setup.dwt**

2. Draw the simple floorplan shown below.

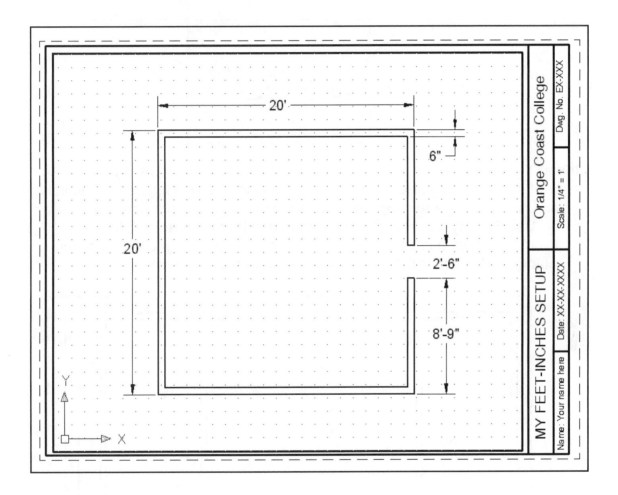

3. Save this drawing as: **EX-10A step 1**

4. Step 2 is on the next page.

Step 2.

5. **Open** the **DesignCenter** (Refer to page 10-2 if necessary.)

6. **Find** your drawing **EX-8C** in Tree View (left side.)

7. Select the **plus box** beside the file to display the contents of **EX-8C**.

8. Select **"Blocks"** to display the blocks in the Content area (right side).

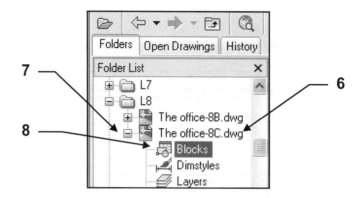

9. **Insert** the Blocks from the Content area to create the drawing below.
    (Refer to page 10-4, How to Insert a Block using DesignCenter, if necessary.)
    a. Use the **drag and drop** method for the blocks that are not rotated.
    b. Use **Specify Coordinate** method for the blocks that are rotated.
    (Note: The Sofa and Chairs are at a 45degree angle but you must use the AutoCAD degree clock to determine the rotation of the block.)

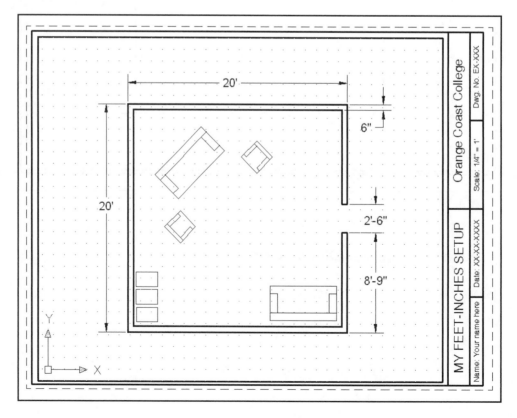

10. **Save as EX-10A step 2.**

EXERCISE 10B

Borrowing settings from another drawing

The following exercise is to demonstrate how quick and easy it is to borrow settings, text styles, dimension styles, layers, layouts, etc. from a previously created drawing file.

1. Using **New** select **acad.dwt**
    Notice this drawing has no layers, layouts, dimension styles, text styles etc.

2. Select the **DesignCenter**

3. Find **My Decimal Setup** in the Tree area.

4. Select the **[+]** beside the file to display Blocks, dimension styles, layouts etc. list.

5. Select **Layer**

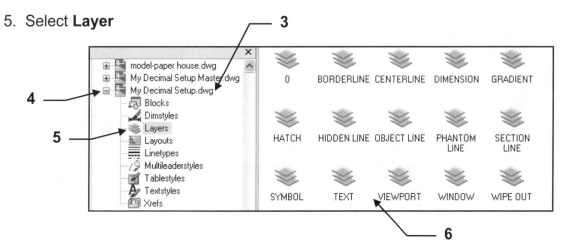

6. Select all of the Layers in the Content area.

7. Click, Drag and drop all of them into the drawing area.
    Now look at your Layers. You should now have more layers than you had previously.

8. Now select Layouts.

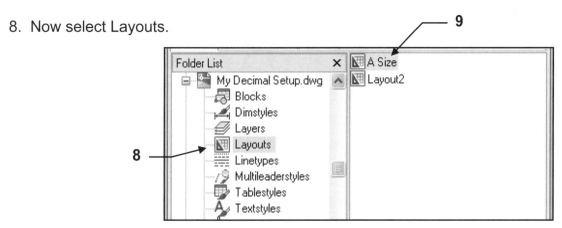

9. Select **A size**

10. Click, drag and drop **A size** into Paper space.
    Notice the **A size** layout tab appears.
    Select it and see what is displayed.
    It should have your Border, Title Block and Viewport.

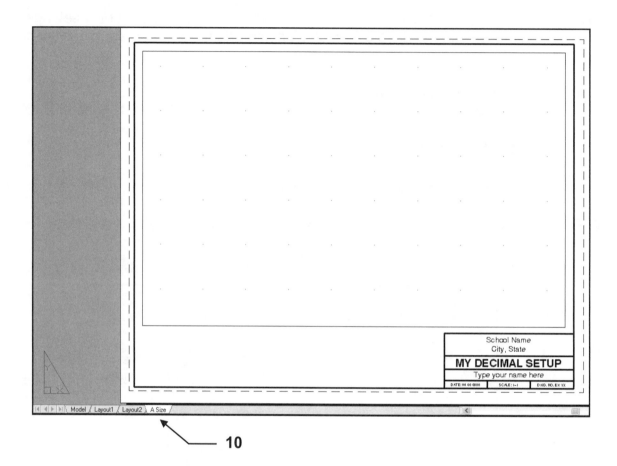

11. Do the same with Dimstyles and Textstyles.

12. **Save as EX-10B**.

LEARNING OBJECTIVES

After completing this lesson, you will be able to:

1. Understand the use of Externally Referenced drawings
2. Insert an Xref drawing.
3. Use the Xref Manager.
4. Bind an Xref to a drawing.
5. Clip an Xref drawing.
6. Edit an Xref drawing
7. Convert an object to a Viewport.
8. Create multiple Viewports and multiple xrefs.
9. Create multiple viewports quick and easy

LESSON 11

External Reference drawings (xref)

The **XREF** command is used to insert an <u>image</u> of another drawing into the current drawing. AutoCAD calls the host drawing the "Parent" and the Xref dwg the "Child". This command is very similar to the INSERT command. But when you externally reference (XREF) the Child dwg into the Parent dwg, the image of the Child dwg appears but only the <u>**path**</u> to where the original Child dwg is stored is loaded into the Parent dwg. The Child dwg does not become a permanent part of the Parent dwg.

Each time you open the Parent dwg, it will look for the Child dwg (via the path) and will load the image of the **current version** of the Child dwg. If the Child dwg has been changed, the new changed version will be displayed. So the Parent dwg will always display the most current version of the Child dwg.

Since only the path information of the Child dwg is stored in the Parent dwg, the amount of data in the Parent dwg does not increase. This is a great advantage.

If you want the Child dwg to actually become part of the Parent dwg, you can **BIND** the Child to the Parent. (Refer to page 11-6)

If you want the image of the Child dwg to disappear temporarily, you can **UNLOAD** it. To make it reappear simply **RELOAD** it. (Refer to page 11-5 & 6)

If you want to delete the image and the path to the Child dwg, you can **DETACH** it. (Refer to page 11-5)

Examples of how the XREF command could be useful?

1. **If you need to draw an elevation**. You could XREF a floor plan and use it basically as a template for the wall, window and door locations. You can get all of the dimensions directly from the floor plan by snapping to the objects. It will not be necessary to refer back and forth between drawings for measurements. If there are any changes to the floor plan at a later date, when you re-open the elevation drawing, the latest floor plan design will appear automatically. To make the floor plan drawing become invisible, simply UNLOAD it. When you need it again, RELOAD it.

2. **If you want to plot multiple drawings on one sheet of paper.** Many projects require "standards" or "detail" drawings that are included in every set of drawings you produce. Currently you probably drew each detail on the original drawing package, or created those transparent "sticky backs". Now, using the XREF command, you would do the following. 1. Make individual drawings of each standard or detail. 2. XREF the standard or detail drawing into the original drawing package. This would not only decrease the size of your file but if you made a change to the standard or detail, the latest revision would automatically be loaded each time you opened the drawing package.

3. **Working with a team of drafters all working on a segment of the project.** (This process works best if your office is networked.) Let's say you are responsible for the furniture layout on an architectural project and the floor plan is not quite finished. You could XREF the current version of the floor plan and start working on the furniture layout. Every couple of hours you can RELOAD the XREF floor plan and your drawing will be automatically updated to the latest work saved on the floor plan.

HOW TO INSERT AN XREF DRAWING

1. Open the Parent drawing. (For example, open My Decimal Setup)

2. Select Layer **Xref**. (I suggest you create a layer for Xref drawings to reside on)

3. Select the **Model** tab
   Note: You should always insert an Xref into Model Space not Paper Space. The viewport can be locked or unlocked, it does not matter.

4. Select the XREF command using one of the following:

Ribbon = Insert tab / Reference panel /

Keyboard = xattach

The following "Select Reference File" dialog box should appear.

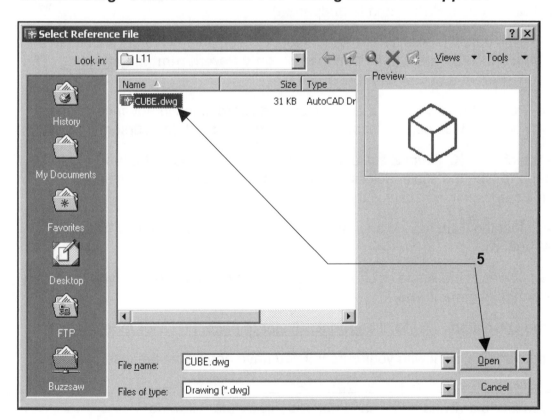

5. Select the file you wish to **XREF** and select **Open.**

continue on the next page….

*The following "**External Reference**" dialog box should appear.*

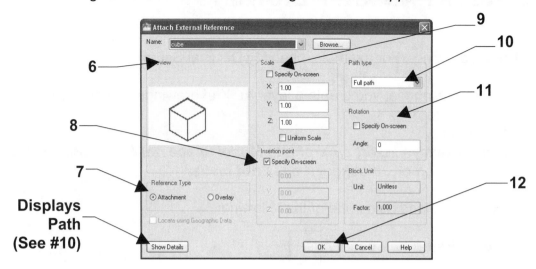

6. The name of the file you selected should appear in the Name box.

7. Select **"Reference Type"**.

   Attachment - This option attaches one drawing to another. If you attach the child to the Parent then attach the Parent to a Grandparent, the **Child remains attached**. (Daisy Chain of Child, Parent and Grandparent)

   Overlay - This option is exactly like Attachment except, if you Overlay a Child to a Parent, then attach a Parent to a Grandparent, **only Parent remains attached**. The Child is not. (Parent and Grandparent only)

8. **Insertion Point -** Specify the X, Y and Z insertion coordinates in advance or check the box "Specify on-Screen" to locate the insertion point with the cursor.

9. **Scale -** Specify the X, Y and Z scale factors in advance or check the box "Specify on-screen" to type the scale factors on the command line. You may select the **Uniform Scale** box if the scale is the same for X, Y and Z.

10. **Path -** Select the Path type to save with the xref. (To see path select "Show Details" button)

11. **Rotation -** Specify the Angle or check the box "Specify on-screen" to type the scale factors on the command line.

12. Select the **OK** button. (Xref will be **faded**. Refer to page 11-12 for fade control.)

13. Place the insertion point with your cursor if you have not preset the insertion point.

Note: If you xref a drawing and it does not appear, try the following:
1. Verify that you xref-ed the drawing into Model space, not paper space.
2. Use View / Zoom / Extents to find the drawing in a viewport, adjust the scale and re-lock the viewport.
3. Go back to the original drawing to verify that it was drawn in model space. Objects drawn in paper space will not xref.
4. Go back to the original drawing and verify that the layers are not frozen.

EXTERNAL REFERENCE PALETTE

The **External Reference Palette** allows you to view and manage the xref drawing information in the current drawing.

Select the External Reference Palette using one of the following:

Ribbon = Insert tab / Reference panel / ⌐

Keyboard = XR

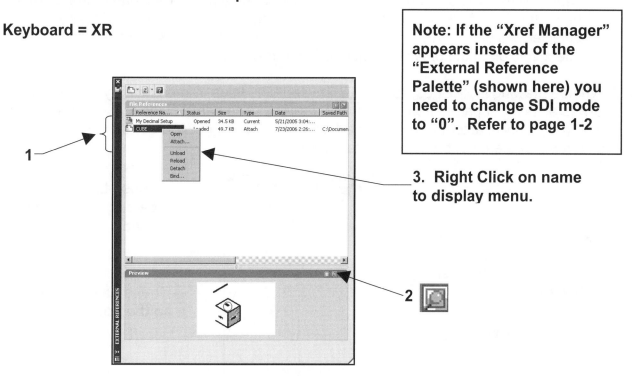

Note: If the "Xref Manager" appears instead of the "External Reference Palette" (shown here) you need to change SDI mode to "0". Refer to page 1-2

3. Right Click on name to display menu.

1. **Title Area** - This describes each drawing in the current drawing.
   Reference Name = The original drawing file name of the drawing
   Status = This lists whether the drawing was opened, loaded, not found or unloaded.
   Size = Indicates the size of the drawing.
   Type = Lists whether the xref drawing is an Attachment or Overlay.
   Date = Lists when the xref drawing was originally referenced.
   Saved Path = This is the path the computer will follow to find the xref drawing listed and load it each time you open the current drawing.

2. **Preview button:** - Displays a preview of the drawing selected.

3. **MENU (Right click on name to display menu)**

   Open - Opens the selected xref for editing in a new window. The new window is displayed after the External Reference Palette is closed.

   Attach - Takes you back to the External Reference dialog box so you can select another drawing to xref.

   Unload - <u>Unload is not the same as Detach</u>. An unloaded xref drawing is not visible but the information about it remains and it can be reloaded at any time.

Reload - This option will reload an unloaded xref drawing or update it. If another team member is working on the drawing and you would like the latest version, select the Reload option and the latest version will load into the current drawing.

Detach - This will remove all information about the selected xref drawing from the current file. The xref drawing will disappear immediately. <u>Not the same as Unload.</u>

Bind – When you bind an xref drawing, it becomes a permanent part of the current base drawing. All information in the Xref manager concerning the xref drawing selected will disappear. When you open the base drawing it will not search for the latest version of the previously inserted xref drawing.

When a drawing is XREFed, its Blocks, Layers, Linetypes, Dimension Styles and Text Styles are kept separate from the current drawing. The name of the xref drawing and a pipe (I) symbol is automatically inserted as a prefix to the newly inserted xref layer. This assures that there will be no duplicate layer names.

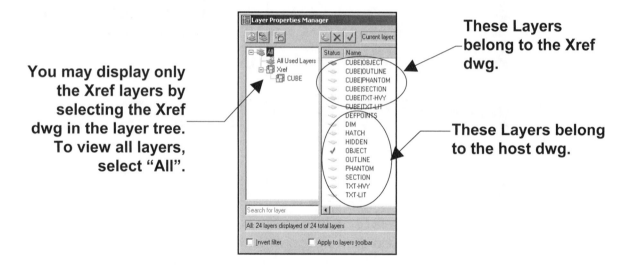

You may display only the Xref layers by selecting the Xref dwg in the layer tree. To view all layers, select "All".

These Layers belong to the Xref dwg.

These Layers belong to the host dwg.

XBIND

It is important to understand that these new names are listed but they cannot be used. Not unless you bind the entire xref drawing or use the Xbind command to bind individual objects.

The following is an example of the Xbind command with an individual layer. You may also use this command for Blocks, Linetypes, Dimension Styles and Text Styles.

1. Type **XBIND** <enter> at the command line.

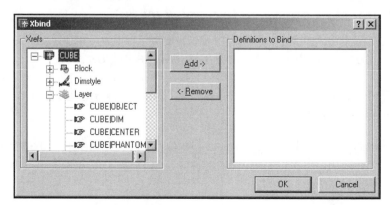

2. Expand the Xref file information list by clicking on the **+** sign beside the Xref drawing file name; then click on the **+** sign beside "Layer". The + sign will then change to a - sign, as shown below.

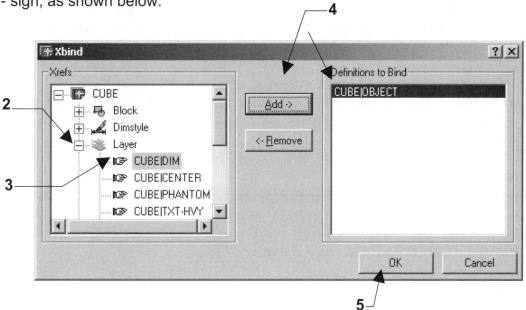

3. Select the layer name to bind.

4. Select the **ADD** button. The xref layer will now appear in the "Definitions to Bind" window.

5. Select the **OK** button.

6. Now select Format / Layers and look at the xref layer. The Pipe (I) symbol between the xref name and the layer name has now changed to 0.
This layer is now usable.

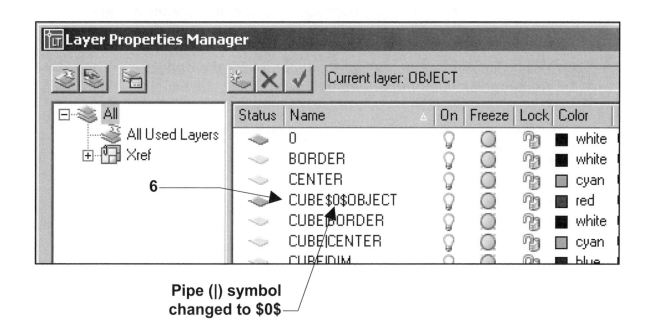

Pipe (|) symbol changed to 0

Clipping an External Reference (Not available in LT)

When you Xref a drawing, sometimes you do not want the entire drawing visible. You may only want a portion of the drawing visible. For example, you may Xref an entire floor plan but actually only want the kitchen area visible within the viewport.

This can be easily accomplished with the Xclip command. After you have inserted an External Referenced drawing, select the Xclip command. You will be prompted to specify the area to clip by placing a window around the area. **All objects outside of the window will disappear.**

How to use the **Xclip** command

1. Select the **Xclip** command using one of the following:

    Ribbon = Insert tab / Reference panel /

    Keyboard = XC

2. _clip Select objects to clip: *select the xref you want to clip <enter>*

    a. Enter clipping option
        [ON/OFF/Clipdepth/Delete/generate Polyline/New boundary] <New>: *N <enter>*

    b. Outside mode - Objects outside boundary will be hidden.
        Specify clipping boundary or select Invert option:
        [Select polyline/Polygonal/Rectangular/Invert clip] <Rectangular>: *R <enter>*

    c. Specify first corner: *select the location for the first corner of the "Rectangular" clipping boundary.*

    d. Specify opposite corner: *select the location for the opposite corner of the "Rectangular" clipping boundary.*

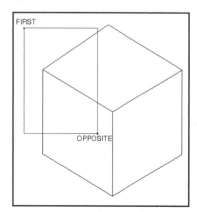
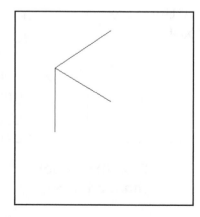

Note: If you select "Invert clip" in step b above the objects inside the boundary will disappear instead of outside the boundary.

CLIPPING OPTIONS

1. Select the XCLIP command.
2. Click on the clipped image.
The following is a description for each option.

Enter clipping option
[ON/OFF/Clipdepth/Delete/generate Polyline/New boundary] <New>:

The description for the options are shown below:

ON / OFF – If you have "clipped" a drawing, you can make the clipped area visible again using the "OFF" option. To make the clipped area invisible again, select the "ON" option.

CLIPDEPTH – This option allows you to select a front and back clipping plane to be defined on a 3D model.

DELETE – To remove a clipping boundary completely.

GENERATE POLYLINE – This option creates a polyline object to represent the clipping border of the selected Xref. (The boundary is invisible by default)

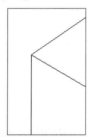

NEW BOUNDARY – Allows you to select a new boundary.

If you select New Boundary option the following prompt appears because you can not create a new boundary unless the old boundary is deleted.

Delete old boundaries, yes or no: <yes>:

The descriptions for the options are shown below:

SELECT POLYLINE – Allows you to select an existing polyline as a boundary.

POLYGONAL –Allows you to draw and irregular polygon, with unlimited corners, as a boundary.

RECTANGULAR – Allows a rectangular shape, with only 2 corners, for the boundary.

INVERT CLIP - Inverts the mode of the clipping boundary: either the objects outside the boundary (default) or inside the boundary are hidden.

EDIT AN EXTERNAL REFERENCED DRAWING

An Xref drawing can be edited very easily using the open command in the Xref palette. The open command opens the Xref drawing (the child) in a separate window, you make the changes and save those changes to the original of the Xref (Child) drawing. Select "Reload" to view the changes in the Parent drawing.

How to use the open command.

1. Open the drawing that contains the External Referenced drawing.

2. Type **XR**

3. Right click on the Xref drawing to be changed.

4. Select **Open** from the short cut menu.

    The selected Xref original drawing will open in a separate window.
    (Note: the "Single-drawing compatibility mode" must be "off".
    Ref. page 1-2)

5. Make the necessary changes.

6. Save the drawing. *(Note: you must use the same name.)*

7. Close the drawing.

    The Parent drawing will reappear with a "balloon message" notifying you that the Xref drawing has been changed and it needs to be "Reloaded".

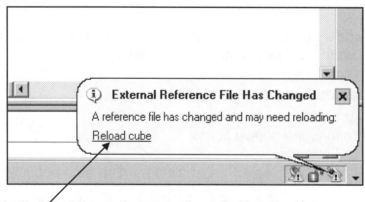

8. Click on the drawing name (underlined and blue)

Note: If the Change Balloon does not appear select <u>Insert / External Reference</u>... See next page.

***If the "Change Balloon" did not appear select Insert / External Reference...
the following dialog box should appear.***

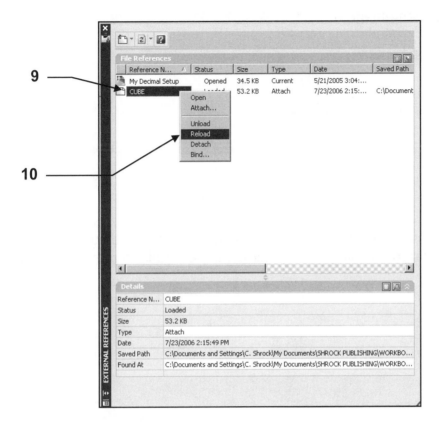

9. Select the drawing to reload.

10. Right click on the name and select the ***Reload*** from the drop down menu.

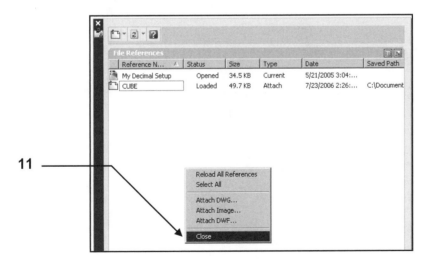

11. Right click again and select ***Close.***

CONTROL THE XREF IMAGE FADE

When you insert an externally referenced drawing the image appears faded. This is merely an aid to easily identify the xref inserted from the parent drawing.

You can control the percentage of fade.

1. Type **xdwgfadectl** <enter>

2. Enter a positive number from 0 to 90.
    This number is a percentage of fade.

    0 = no fade
    90 = maximum fade
    70 = initial AutoCAD setting

    Note:
    Fade control is saved to your computer. It is not saved with the individual file.
    If you open the parent drawing on another computer the fade may appear differently.

Convert an Object to a Viewport

You may wish to have a viewport with a shape other than rectangular. AutoCAD allows
you to convert Circles, Rectangles, Polygons, Ellipse and Closed Polylines to a Viewport.

1. You must be in Paperspace.
2. Draw the shape of the viewport using one of the objects listed above.

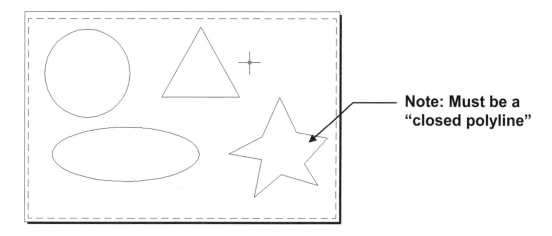

**Note: Must be a
"closed polyline"**

3. Type **MV <enter>**

4. Specify corner of viewport or
    [ON/OFF/Fit/Shadeplot/Lock/Object/Polygonal/Restore/LAyer/2/3/4] <Fit>: **O <enter>**

5. Select object to clip viewport: **select object to convert**

    Select object to clip viewport: Regenerating model.

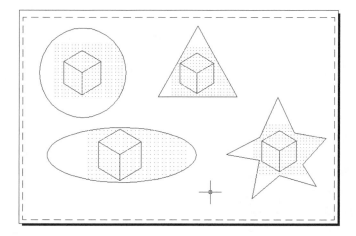

Now you can see through to model space. This is like looking out all the windows in your
living room and seeing the same tree growing in your front yard.

Creating Multiple Viewport and Multiple Xrefs

When using Xrefs you may to XREF more than one drawing. And you will also require multiple viewports. Multiple viewports are easy but you will have to think a little about multiple xrefs.

The following is an example of how to control the viewing of multiple xrefs. The actual exercise, with step-by-step instructions, is EX-11A

1. Open **My Decimal Setup**

2. Select the **A Size** layout tab.

3. Delete any viewports that already exist. (Click on each frame then select Erase)

4. Select the viewport layer.

5. Create 2 viewports approximately as shown below.
   First one, than the other.
   Simple so far, huh?

6. Double click inside the Viewport A, on the left.

7. Select the XREF layer.

8. Xref a drawing into Viewport A. (7C would be a good one to use)

9. Select Zoom / Extents (You should see the entire drawing inside Viewport A)

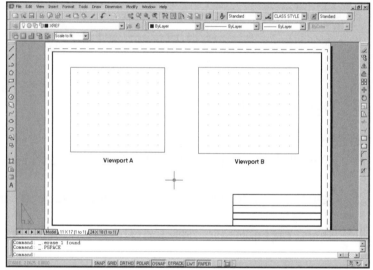

10. Now activate Viewport B (double click inside Viewport B)

11. Select Zoom / Extents (You should see the same Xref drawing inside Viewport B also)

Now take a minute to think about this. This is an important concept to understand. Pretend that the Viewport frames are 2 windows in your living room and the xref drawing is a tree in your front yard. If you stand infront of one of the windows you can see the tree. Then if you walk to the other window you see the same tree. It is the same tree, right?

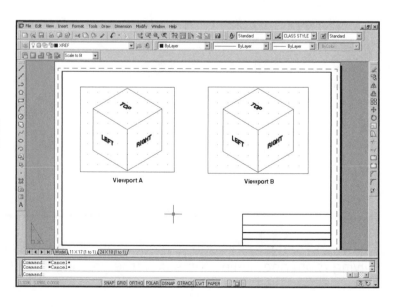

This is basically what is happening with AutoCAD. You have one xref drawing in Model Space (your front yard) and you are in Paper Space (your living room) looking through the viewports (your windows) to Model Space.

Now let's make the xref drawing in Viewport B disappear.

If you **do not** want to see the xref drawing in Viewport B you <u>must "**Freeze in Current Viewport**" the layers of the xrefed drawing inside Viewport B.</u>
This is easier than is sounds.

12. Activate Viewport B. (click in it)

13. Select the **Layer Properties Manager**

Scroll up and down and look at the layer names. Notice some layers have been added. (Refer to page 11-6)

When the drawing was Xrefed, the layers came with that drawing.

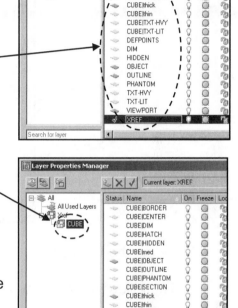

14. To view only the layers that belong to the Xref drawing select the xref drawing name in the **Filter Tree** area.

15. Select all of the layers that belong to the XREF drawing. (highlight them)

16. Select the "**Freeze in Current Viewport**" box. Now this is very important: **Do not select "Freeze"**. Slide all the way over to the right hand side of the list of layers and select the "**Freeze in Current Viewport**" column. (Displayed as: **VP Freeze**)

17. Close the dialogue box.

18. The Xrefed drawing should have disappeared in Viewport B. If it didn't do items 12 through 16 again.

19. Now you should adjust the scale in Viewport A and lock it.

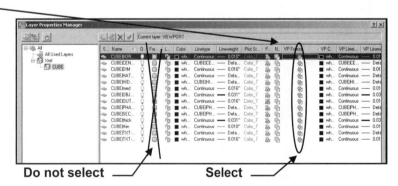

Do not select ———   Select ———

Now think about this. If you were to Xref another different drawing into Viewport B, it would also be visible in Viewport A (just another tree in your front yard). So you would have to activate Viewport A, select the layers that belong to the newly Xrefed drawing and select "Freeze layers in Current Viewport". Then only one drawing would be visible in each viewport.

CREATING MULTIPLE VIEWPORTS – A QUICK METHOD

1. Select **VIEW tab / VIEWPORTS panel /** 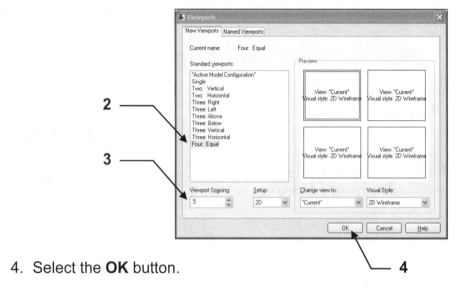 New

2. Select **Four: Equal** from the column on the left.
   This will divide the paperspace into 4 equal viewports

3. Enter **.50 Viewport Spacing**
   This the spacing between each of the new viewports.

4. Select the **OK** button.

5. The following prompt will appear:

   Specify first corner or [Fit] <Fit>: *Enter the location of the first corner of the area to divide and then the opposite corner.*

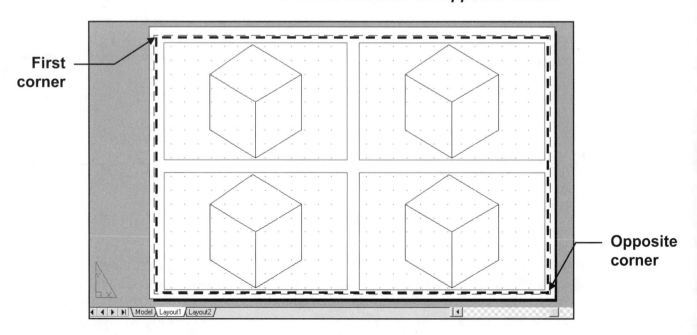

Note: If you have objects in Modelspace, they will appear in each viewport.

EXERCISE 11A

XREF MULTIPLE DRAWINGS

1. Select your **My Decimal Set Up** template.

2. Select the "model" tab and **draw** a Rectangle (3.50" L X 2" W) and **save** as **"RECT"**

3. **Close** the drawing.

4. **Select** My Decimal Set Up template **again**.

5. Select the "model" tab and **Draw** a Circle (1.50" Radius) and **save** as **"CIRCLE"**.

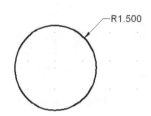

6. **Close** the drawing.

7. **Select** My Decimal Setup template **again.**

8. Select the **A Size** tab.

9. Erase the existing Viewport frame. (Click on it, select erase and <enter>)

10. Select layer Viewport.

11. Create 2 new Viewports as follows:
    a. Draw 2 circles, **R2.50**, as shown below.
    b. Select the "**MV**" command and convert them to Viewports. (Refer to page 11-13)

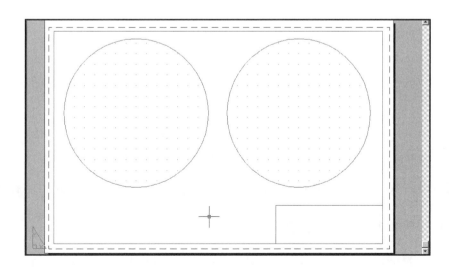

12. **Add Viewport Titles** as shown below. (**In paper space**)
   a. Use "Single Line Text"
   b. Style = Text Classic
   c. Height = .35
   d. Layer = Text

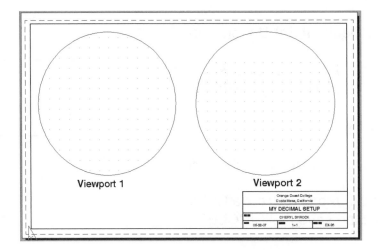

13. Double click inside of Viewport 1 to **activate Model space**.

14. Change to the XREF layer.

15. **Xref** the **Rect** drawing, you made previously, as follows:
   a. Select the Attach command **(Refer to page 11-3)**
   b. Find and Select **Rect** drawing.
   c. Select the **OPEN** button.
   d. Select the **OK** button.
   e. Click inside Viewport 1.

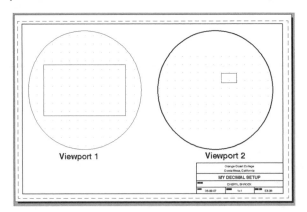

16. Use **Zoom / All** or **Extents** inside each viewport so you can see the rectangle.

17. **Adjust the scale** in Viewport 1 to **1 : 1.**

18. **"Pan"** the **Rect** drawing inside of Viewport 1 to display it as shown.
   (Do not use the Zoom commands or you will have to re-adjust the scale)

Now you are going to make the Rect drawing disappear in Viewport 2.

19. Freeze all the layers that belong to **Rect** drawing in Viewport 2 as follows:
    a. **Activate** the Viewport 2.
    b. Select **LAYER Properties Manager (type LA)**
    c. Select **Rect** in the Filter tree.
    d. Select all the layers that begin with **Rect.**

e. Select the **"VP Freeze"** column.
f. Close the Layer Properties Manager.

The Rect drawing inside Viewport 2 should have disappeared.

The next step is to XREF a second drawing into the other viewport.
Then you will make the newly xref-ed drawing disappear in Viewport 1.
When you have completed step 24 you should see only the Rectangle in Viewport 1
and only the Circle in Viewport 2.

20. **Xref** drawing **Circle** into the **Viewport 2** as follows:
    a. Select the **XREF** layer.
    b. Select the **Attach** command. (See page 11-3)
    c. Find and Select the **Circle** drawing.
    d. Select the **OPEN** button.
    e. Select the **OK** button.
    f. Click inside Viewport 2.

21. Use **Zoom / All** or **Extents** inside Viewport 2 if you can't see the Circle drawing.

22. **Adjust the scale** in Viewport 2 to **1 : 1.**

23. **"Pan"** the **Circle** drawing inside of Viewport 2 to display it as shown on next page.
    (Do not use the Zoom commands or you will have to re-adjust the scale)

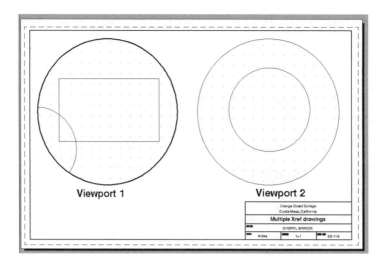

Viewport 1   Viewport 2

Orange Coast College
Costa Mesa, California
Multiple Xref drawings
CHERYL SHROCK
today   1=1   EX-11A

Now you will make the Circle drawing disappear in Viewport 1.

24. "Freeze in Current Viewport" all the layers that belong to the Circle drawing Viewport 1 as follows:
   a. **Activate** Viewport 1.
   b. Select **LAYER Properties Manager (Type LA)**
   c. Select the **Circle** in the **Filter tree** area.
   d. Select all the layers that begin with **Circle**.
   e. Select the "**VP Freeze**" column.
   f. Close the Layer Properties Manager.

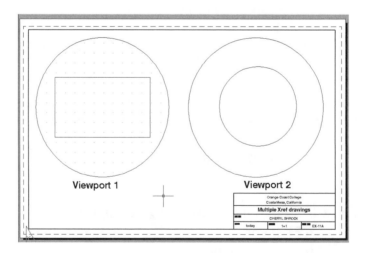

Viewport 1   Viewport 2

Orange Coast College
Costa Mesa, California
Multiple Xref drawings
CHERYL SHROCK
today   1=1   EX-11A

The Xref drawing, Circle, should have disappeared in Viewport 1.

25. Save as **EX-11A**.

26. **Plot** using Plot Page Set up **Plot Setup A.**

EXERCISE 11B

Creating Multi-scaled Views

1. Select your **My Decimal Setup** template.
2. Select the **A Size** tab.
3. Important: Erase the existing Viewport frame. (Click on it, select erase and <enter>)
4. Select the Viewport layer.
5. Create 3 new Viewports as shown below using the **Three: Right** (see page 11-16).

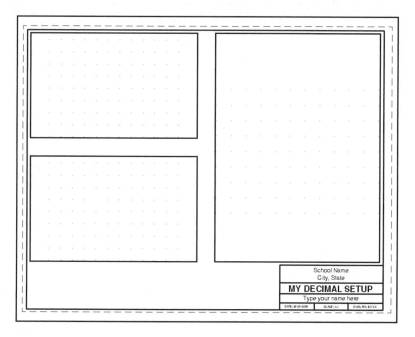

6. Add Viewport Titles as shown below. (In paperspace)
   a. Use "Single Line Text"
   b. Style = Text-Classic
   c. Height = .25
   d. Layer = Text

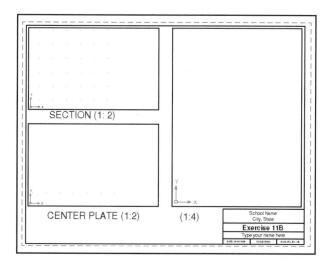

7. Double click inside one of the Viewports to **activate Model space**.

8. **Xref** drawing 6A as follows:
   a. Select the **Attach** command
   b. Find and Select 6A
   c. Select the **OPEN** button.
   d. Remove the check mark from all of the "Specify On-Screen" boxes.
   e. Select the **OK** button.

9. **Adjust the scale** in each Viewport to the scale listed under each viewport as shown below.

10. **"Pan"** the drawing inside each viewport to display the correct area. (Do not use the Zoom commands or you will have to re-adjust the scale)

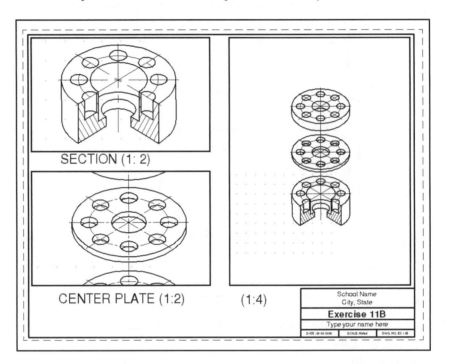

11. Save as **EX-11B**

12. **Plot** using Plot Page Set up **Plot Setup A**

EXERCISE 11C

CLIPPING AN EXTERNAL REFERENCE

1. Open **EX-11B.**
2. Select the **Model** tab.

Note: If the drawing does not appear when you select the Model tab, type: Regen <enter> and it will appear.

3. Select **XCLIP** command. (Ref to page 11-8)
4. Clip the lower section to appear as **shown below**:
   a. Select objects.
   b. Select <u>New Boundary</u> and <u>Rectangular</u>

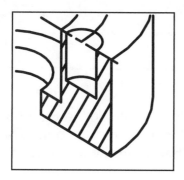

5. **Save as EX-11C1**

6. Select the **XCLIP** command <u>again</u>.
   a. Select Objects.
   b. Select **OFF**
   The entire drawing should reappear.

7. Select the **XCLIP** command <u>again</u>.
8. Clip the lower section to appear approximately as shown **on the right**:
   a. Use <u>New Boundary</u> .
   b. Select "<u>Yes</u>" when prompted to "<u>Delete old boundary</u>"
   c. Use <u>Polygonal.</u>

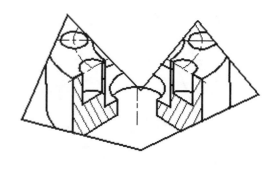

9. **Save as EX-11C2**

MISSING XREF DRAWINGS

When you xref a drawing AutoCAD saves the path to that drawing. So everytime you open the master drawing AutoCAD looks for the xref using that path.

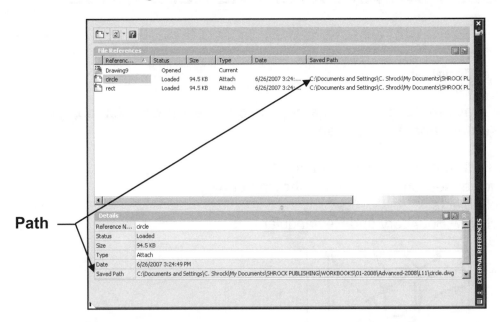

Path

If you move the drawing and you <u>do not change</u> the path, in the External Reference palette, AutoCAD will not be able to find it. If AutoCAD can't find the xref it will display the "old path" and a box will appear asking "**What do you want to do?**".

Example:

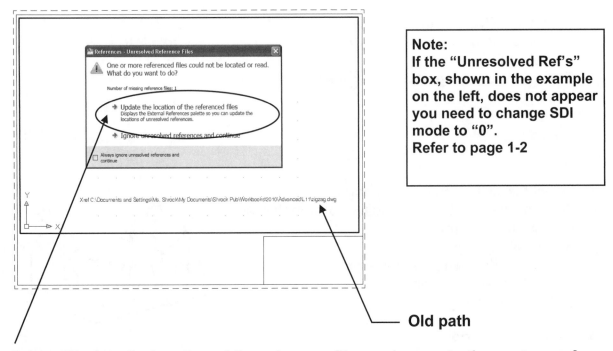

Note:
If the "Unresolved Ref's" box, shown in the example on the left, does not appear you need to change SDI mode to "0".
Refer to page 1-2

Old path

Select "Update the location of the reference files and go on to the next page for "How to change the path".

How to change the path

1. Select the file.

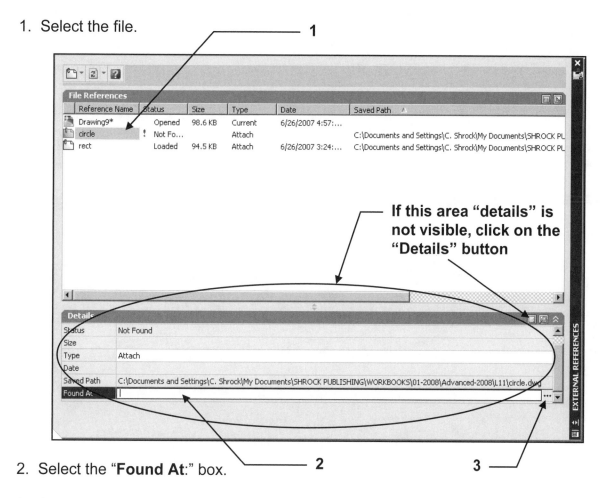

If this area "details" is not visible, click on the "Details" button

2. Select the "**Found At**:" box.

3. Select the **Browse** button.

4. Locate the file and select **Open**.
   The Path will update and the drawing file will automatically load.

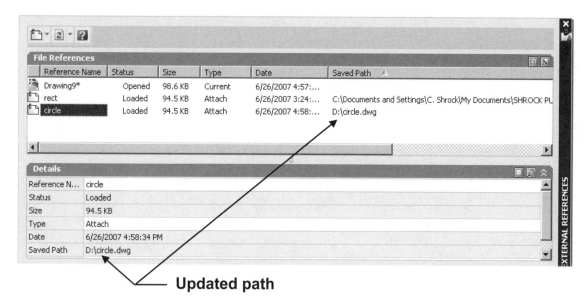

Updated path

NOTES:

LEARNING OBJECTIVES

After completing this lesson, you will be able to:

1. Dimension using Datum dimensioning.
2. Use alternate dimensioning for inches and M illimeters.
3. Assign Tolerances to a part.
4. Use Geometric tolerances.
5. Typing Geometric Symbols.

LESSON 12

ORDINATE Dimensioning

Ordinate dimensioning is primarily used by the sheet metal industry. But many others are realizing the speed and tidiness this dimensioning process allows.

Ordinate dimensioning is used when the X and the Y coordinates, from one location, are the only dimensions necessary. Usually the part has a uniform thickness, such as a flat plate with holes drilled into it. The dimensions to each feature, such as a hole, originate from one "datum" location. This is similar to "baseline" dimensioning. Ordinate dimensions have only one datum. The datum location is usually the lower left corner of the object.

Ordinate dimensions appearance is also different. Each **dimension** has only one **leader line** and a **numerical value**. Ordinate dimensions do not have extension lines or arrows.

Example of Ordinate dimensioning:

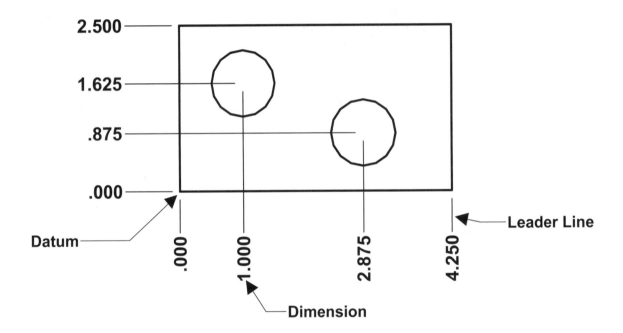

Note:
Ordinate dimensions can be Associative and are Trans-spatial. Which means that you can dimension in paperspace and the ordinate dimensions will remain associated to the object they dimension. (Except for Qdim ordinate)

Refer to the next page for step by step instructions to create Ordinate dimensions.

Creating Ordinate dimensions

1. Move the "Origin" to the desired "datum" location as follows:
   Note: This must be done in Model Space.

2. Select the Ordinate command using one of the following:

   Ribbon = Annotate tab / Dimension panel /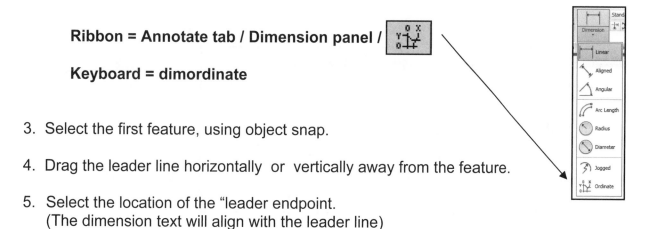

   Keyboard = dimordinate

3. Select the first feature, using object snap.

4. Drag the leader line horizontally or vertically away from the feature.

5. Select the location of the "leader endpoint.
   (The dimension text will align with the leader line)

Use "Ortho" to keep the leader lines straight.

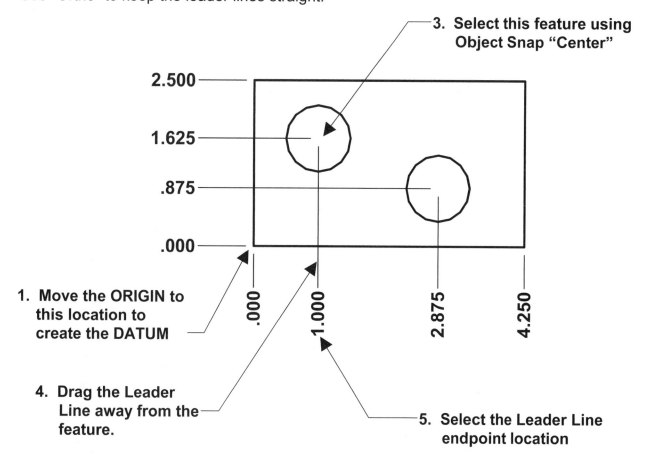

3. Select this feature using Object Snap "Center"

1. Move the ORIGIN to this location to create the DATUM

4. Drag the Leader Line away from the feature.

5. Select the Leader Line endpoint location

JOG an Ordinate dimension

If there is insufficient room for a dimension you may want to jog the dimension.
To **"jog"** the dimension, as shown below, turn "**Ortho**" **off** before placing the Leader Line endpoint location. The leader line will automatically jog. With Ortho off, you can only indicate the feature location and the leader line endpoint location, the leader line will jog the way it wants to.

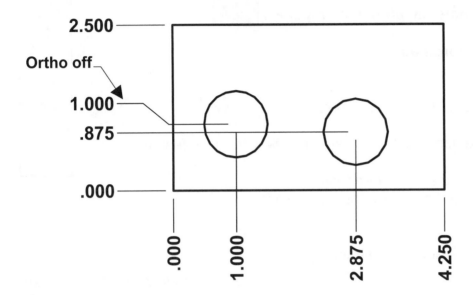

Quick Dimension with Ordinate dimensioning

(Not available in version LT)

1. Select **DIMENSION / Quick Dimension** (Refer to Beginning Workbook, lesson 20)
2. Select the geometry to dimension <enter>
3. Type **"O"** <enter> to select Ordinate
4. Type **"P"** <enter> to select the **datumPoint** option.
5. Select the datum location on the object. (use Object snap)
6. Drag the dimensions to the desired distance away from the object.

> **Note: Qdim can be associative but is not trans-spatial.**
> **If the object is in Model Space, you must dimension in Model Space.**

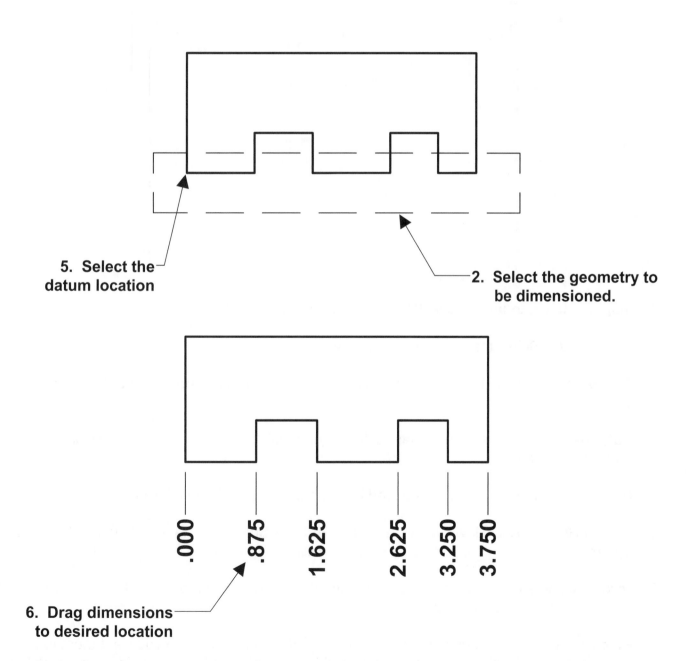

5. Select the datum location

2. Select the geometry to be dimensioned.

6. Drag dimensions to desired location

.000 .875 1.625 2.625 3.250 3.750

ALTERNATE UNITS

The options in this tab allow you to display inches as the primary units and the millimeter equivalent as alternate units. The millimeter value will be displayed inside brackets immediately following the inch dimension. Example: 1.00 [25.40]

1. Select a **DIMENSION STYLE** then **Modify** button.
2. Select the **ALTERNATE UNITS** tab.

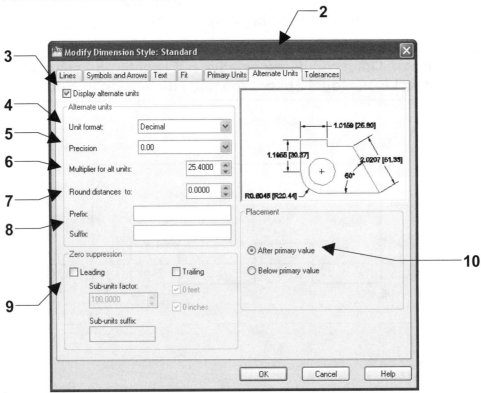

3. **Display alternate units.** Check this box to turn ON alternate units

4. **Unit format**. Select the Units for the alternate units.

5. **Precision** Select the Precision of the alternate units. This is independent of the Primary Units.

6. **Multiplier for all units** The primary units will be multiplied by this number to display the alternate unit value.

7. **Round distance to** Enter the desired increment to round off the alternate units value.

8. **Prefix / Suffix** This allows you to include a Prefix or Suffix to the alternate units. Such as: type **mm** to the Suffix box to display **mm** (for millimeters) after the alternate units.

9. **Zero Suppression** If you check one or both of these boxes, it means that the zero will not be drawn. It will be suppressed.

10. **Placement** Select the desired placement of the alternate units. Do you want them to follow immediately after the Primary units or do you want the Alternate units to be below the primary units?

TOLERANCES

When you design and dimension a widget, it would be nice if when that widget was made, all of the dimensions were exactly as you had asked. But in reality this is very difficult and or expensive.   So you have to decide what actual dimensions you could live with. Could the widget be just a little bit bigger or smaller and still work? This is why tolerances are used.

A **Tolerance** is a way to communicate, to the person making the widget, how much larger or smaller this widget can be and still be acceptable. In other words each dimension can be given a maximum and minimum size. But the widget must stay within that **"tolerance"** to be correct.   For example: a hole that is dimensioned 1.00 +.06 -.00 means the hole is nominally 1.00 but it can be as large as 1.06 but can not be smaller than 1.00.

1. Select **DIMENSION STYLE / MODIFY**
2. Select the **TOLERANCES UNITS** tab.

Note: if the dimensions in the display look strange, make sure "Alternate Units" are turned OFF.

---

3. **Method**
The options allows you to select how you would like the tolerances displayed. There are 5 methods: None, Symmetrical, Deviation, Limits. (Basic is used in geometric tolerancing and will not be discussed at this time)

Refer to the next page for descriptions of methods.

4. **Scaling for height**. This controls the height of the tolerance text. The entered value is a percentage of the primary text height. If .50 is entered, the tolerance text height will be 50% of the primary text height.

5. **Vertical position**. This controls the placement of the tolerance text in relation to the primary text. The options are Top, Middle and Bottom. Whichever option you select, it will align the tolerance text with the bottom of the primary text.

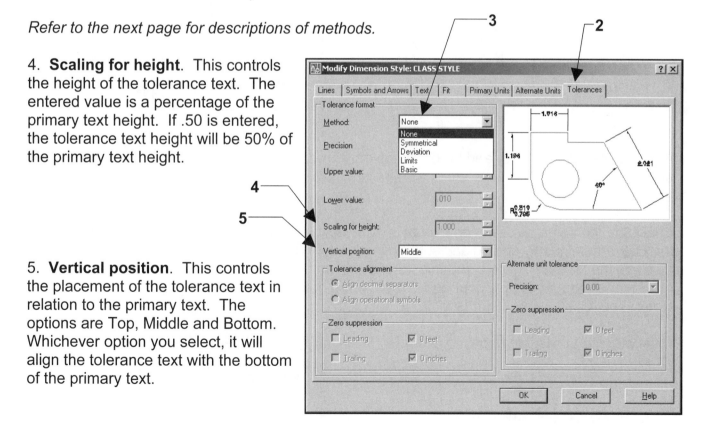

SYMMETRICAL is an equal bilateral tolerance. It can vary as much in the plus as in the negative. Because it is equal in the plus and minus direction, only the "Upper value" box is used. The "Lower value" box is grayed out.

Example of a Symmetrical tolerance :

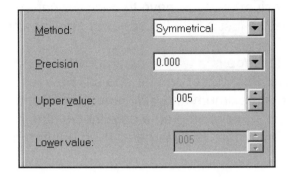

DEVIATION is an unequal bilateral tolerance. The variation in size can be different in both the plus and minus directions. Because it is different in the plus and the minus the "Upper" and "Lower" value boxes can be used.

Example of a Deviation tolerance:

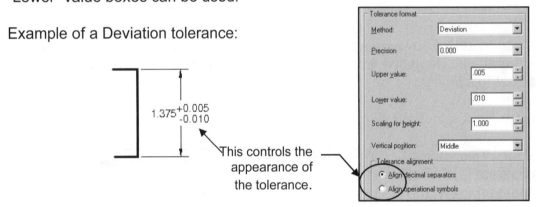

This controls the appearance of the tolerance.

Note: If you set the upper and lower values the same, the tolerance will be displayed as symmetrical.

LIMITS is the same as deviation except in how the tolerance is displayed. Limits calculates the plus and minus by adding and subtracting the tolerances from the nominal dimension and displays the results. Some companies prefer this method because no math is necessary when making the widget. Both "Upper and Lower" value boxes can be used.
Note: The "Scaling for height" should be set to "1".

Example of a Limits tolerance:

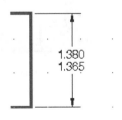

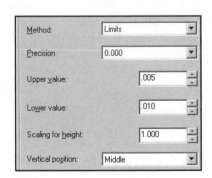

GEOMETRIC TOLERANCING

Geometric tolerancing is a general term that refers to tolerances used to control the form, profile, orientation, runout, and location of features on an object. Geometric tolerancing is primarily used for mechanical design and manufacturing. The instructions below will cover the Tolerance command for creating geometric tolerancing symbols and feature control frames.

If you are not familiar with geometric tolerancing, you may choose to skip this lesson.

1. Select the **TOLERANCE** command using one of the following:

   Ribbon = Annotate tab / Dimension panel ▼ / ⊡.1

   Keyboard = tol

The Geometric Tolerance dialog box, shown below, should appear.

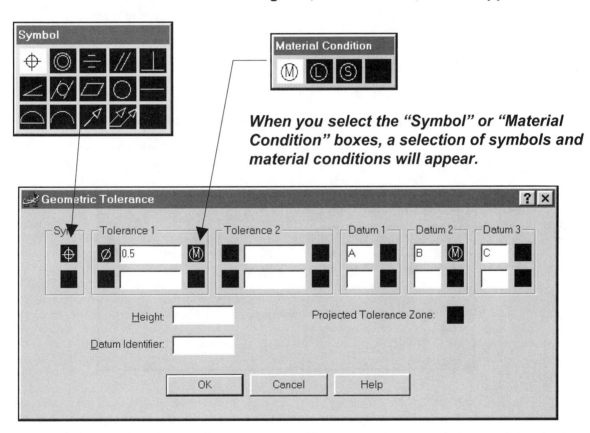

When you select the "Symbol" or "Material Condition" boxes, a selection of symbols and material conditions will appear.

2. Make your selections and fill in the tolerance and datum boxes.
3. Select the **OK** box.
4. The tolerance should appear attached to your cursor. Move the cursor to the desired location and press the left mouse button.

Note: the size of the Feature Control Frame above, is determined by the height of the dimension text.

GEOMETRIC TOLERANCES and QLEADER

The **Qleader** command allows you to draw leader lines and access the dialog boxes used to create feature control frames in one operation. **Do not use Multileader**

1. Type **Leader <enter>**
2. Select **"Settings"**
3. Select the **Annotation** tab
4. Select the **Tolerance** option
5. Select **OK** button

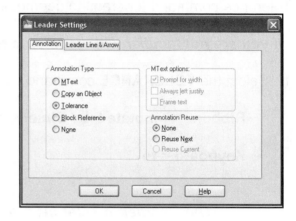

6. Place the first leader point. **P1**
7. Place the next point **P2**

8. Press <enter> to stop placing leader lines

The Geometric Tolerance dialog box will appear.

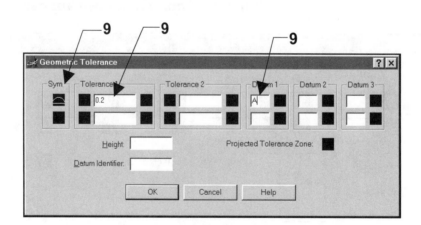

9. Make your selections and fill in the tolerance and datum boxes.

10. Select the **OK** button.

DATUM FEATURE SYMBOL

A datum in a drawing is identified by a **"datum feature symbol"**.

To create a *datum feature symbol*: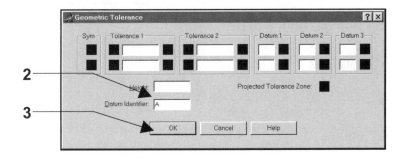

1. Select **Annotate** tab / **Dimension panel** ▼/
2. Type the "**datum reference letter**" in the "**Datum Identifier**" box.
3. Select the **OK** button.

Note: the size is determined by the text height setting within the *current* "dimension style".

To create a *datum feature symbol* combined with a *feature control frame*:

1. Select **Annotate** tab / **Dimension panel** ▼/
2. Make your selections and fill in the tolerance.
3. Type the "**datum reference letter**" in the "**Datum Identifier**" box.
4. Select the **OK** button.

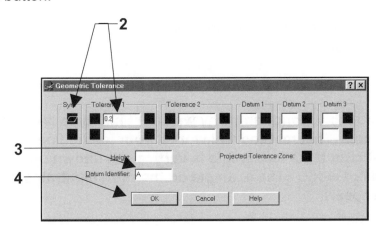

DATUM TRIANGLE

A datum feature symbol, in accordance with ASME Y14.5M-1994, includes a leader line and a datum triangle filled. You can create a wblock or you can use the two step method below using Dimension / Tolerance and Leader.

1. Select **Dimension Style** and then **Override**

2. Change the Leader Arrow to **Datum triangle filled**

3. Select **Set Current** and **close.**

2

4. Type **Leader** <enter>

5. Place the 1st point (the triangle endpoint)

6. Place the 2nd point (Ortho ON)

7. Press **<enter> twice.**

8. Select **None** from the options.

If you were successful,
a datum triangle filled with
a leader line should appear.

6

5

9. Next create a datum feature symbol. (Follow the instructions on the previous page.) **A**

10. Now move the datum feature symbol to the endpoint of the leader line to create the symbol below left.

You are probably wondering why we didn't just type "A" in the identifier box. That method will work if your leader line is horizontal. But if the leader line is vertical, as shown on the left, it will not work. (The example on the right illustrates how it would appear)

TYPING GEOMETRIC SYMBOLS

If you want geometric symbols in the notes that you place on the drawing, you can easily accomplish this using a font named **GDT.SHX**. This font will allow you to type normal letters and geometric symbols, in the same sentence, by merely pressing the SHIFT key when you want a symbol. Note: the CAPS lock must be ON.

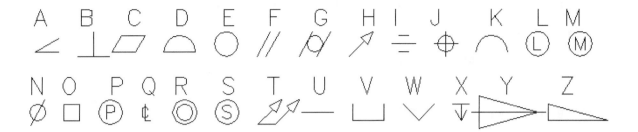

1. First you must create a new text style using the **GDT.SHX** font.

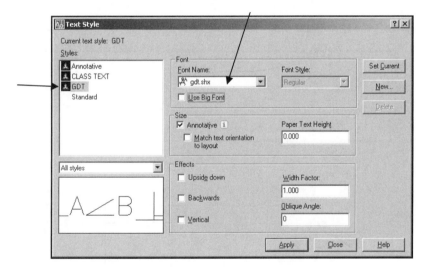

2. **CAPS LOCK** must be **ON**.

3. Select **Single Line or Multiline text**

4. Now type the sentence shown below. When you want to type a symbol, press the **SHIFT** key and type the letter that corresponds to the symbol. For example: If you want the diameter symbol, press the shift key and the "N" key. (Refer to the alphabet of letters and symbols shown above.)

$$3X \; \varnothing.44 \quad \sqcup\varnothing1.06 \; \overline{\vee}.06$$

Can you decipher what it says?
(Drill (3) .44 diameter holes with a 1.06 counterbore diameter .06 deep)

EXERCISE 12A

INSTRUCTIONS:

1. Open **My Decimal Setup.dwt** and select the **A Size** layout tab.

2. **Draw** the objects shown below.

3. Dimension using **Ordinate** dimensioning

4. Viewport scale is 1 : 1

5. **Save** the drawing as: **EX12A**

6. Plot using Plot Page Setup **Plot Setup A**

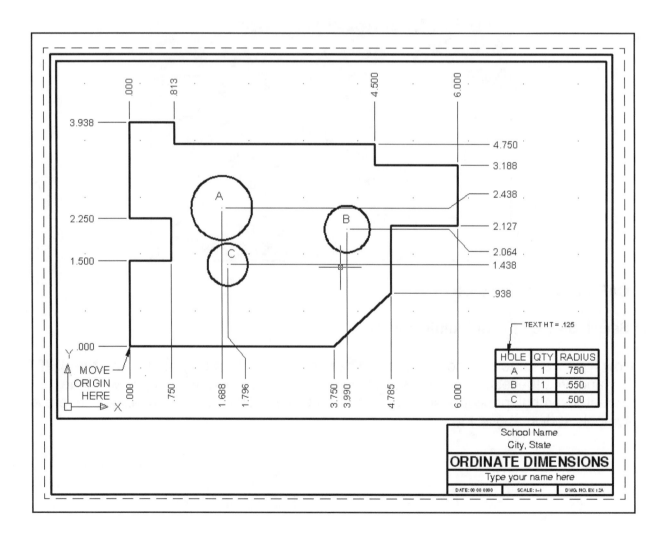

EXERCISE 12B

INSTRUCTIONS:

1. Open **My Decimal Setup.dwt** and select the **A Size** layout tab.

2. **Draw** the objects shown below.

3. Dimension using **Alternate Units**

4. Viewport scale is 1 : 1

5. **Save** the drawing as: **EX12B**

6. Plot using Plot Page Setup **Plot Setup A**

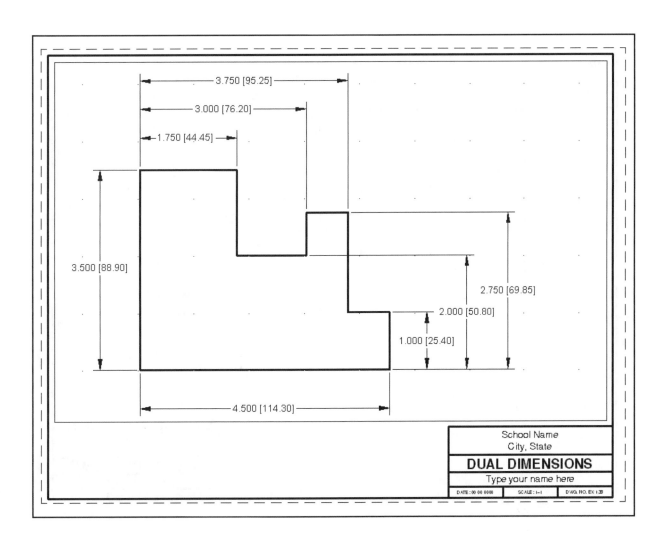

EXERCISE 12C

INSTRUCTIONS:

1. Open **EX-12B** and save immediately as **EX-12C**.

2. Change the dimensions to **Deviation and Symmetrical**. (Use **Properties**)

Note: The Upper and Lower values must be set for each dimension. Or you may use Properties to edit each dimensions. It does get a little tedious.

3. **Save** the drawing as: **EX12C**

4. Plot using Plot Page Setup **Plot Setup A**

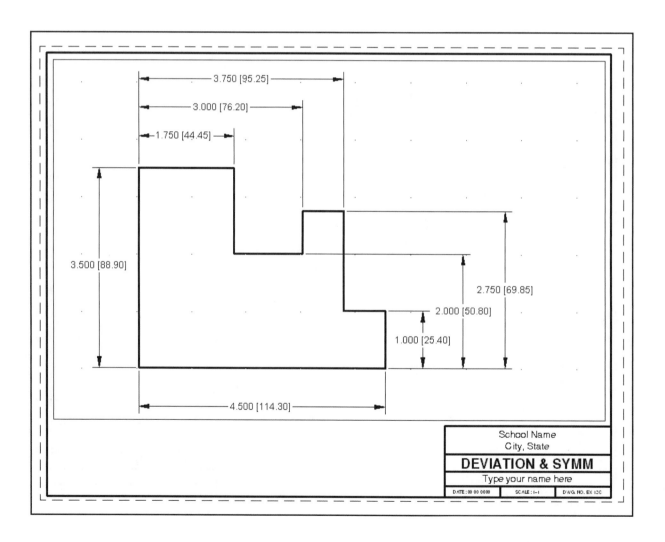

School Name	
City, State	
DEVIATION & SYMM	
Type your name here	

| DATE : 00 00 0000 | SCALE : 1-1 | DWG. NO. EX 12C |

EXERCISE 12D

INSTRUCTIONS:

1. Open **EX-12C** and save immediately as **EX-12D**.

2. Change the dimensions to **Limits**. (Use **Properties**)

Note: The Upper and Lower values must be set for each dimension. Or you may use Properties to edit each dimensions. It does get a little tedious.

3. **Save** the drawing as: **EX12D**

4. Plot using Plot Page Setup **Plot Setup A**

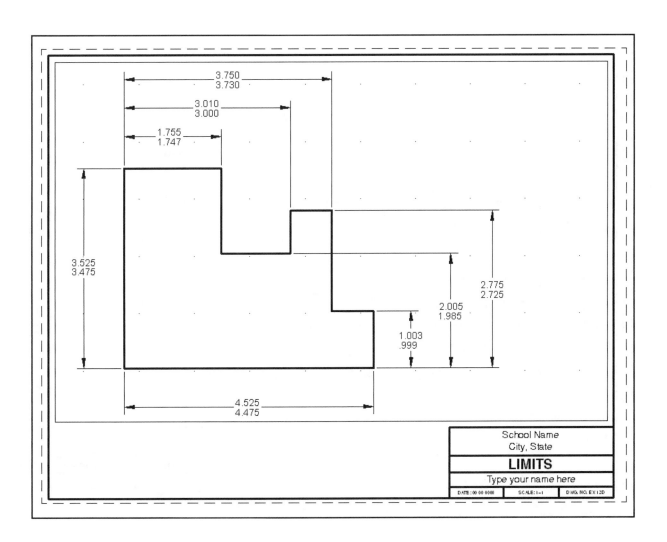

EXERCISE 12E

INSTRUCTIONS:

1. Open **My Decimal Setup.dwt** and select the **A Size** layout tab.

2. **Draw** the objects shown below.

3. Dimension using **Geometric Tolerances** dimensioning

4. Viewport scale is 1 : 1

5. **Save** the drawing as: **EX12E**

6. Plot using Plot Page Setup **Plot Setup A**

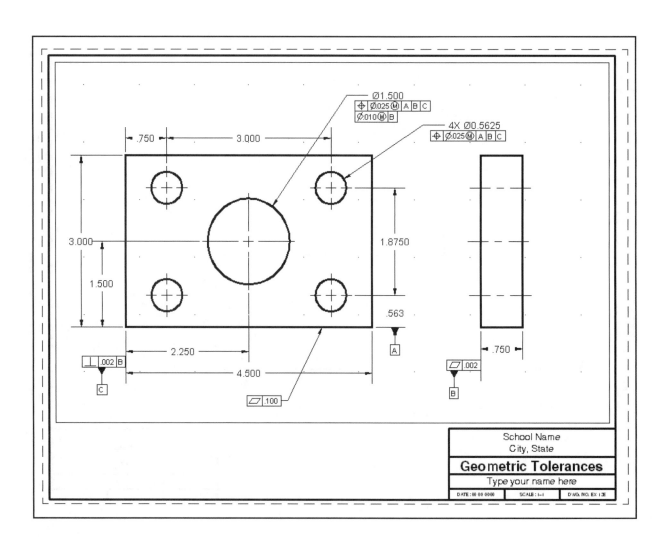

LEARNING OBJECTIVES

After completing this lesson, you will be able to:

1. Understand the concept of 3D
2. Enter the 3D Workspace
3. Manipulate and rotate the 3D Model
4. Understand the difference between:
   Wireframe, Surface and Solid Modeling

The following lessons are an introduction to 3D. These lessons will give you a good understanding of the <u>basics</u> of AutoCAD's 3D program.

Unfortunately, LT users will not be able to do Lessons 13 through 20. The LT version does Not have solid modeling commands.

Note: Increased RAM requirement
The 3D visualization and rendering abilities of AutoCAD requires much more memory than ever before. It will run with 512MB but Autodesk recommends 2GB of RAM and a video card with at least 128MB of its own memory. Check Autodesk's website (www.autodesk.com) for the list of supported video cards.

LESSON 13

INTRODUCTION TO 3D

We live in a three-dimensional world, yet most of our drawings represent only two dimensions. In Lesson 6 you made an isometric drawing that appeared three-dimensional. In reality, it was merely two-dimensional lines drawn on angles to give the appearance of depth.

In the following lessons you will be presented with a basic introduction to AutoCAD's 3D techniques for constructing and manipulating objects. These lessons are designed to give you a good start into the environment of 3D and encourage you to continue your education in the world of CAD.

DIFFERENCES BETWEEN 2D AND 3D

Axes
In 2D drawings you see only one plane. This plane has two axes, X for horizontal and Y for vertical. In 3D drawings an additional plane is added. This third plane is defined by an additional axis named Z. The direction of the positive Z-axis basically comes out of the screen toward you and gives the object height. To draw a 3D object you must input all three, X, Y and Z axes coordinates.

A simple way to visualize these axes is to consider the X and Y axes as the ground and the Z axis as a Tree growing up (positive Z) from the ground or the roots growing down (negative Z) into the ground.

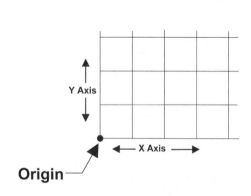

2D Coordinate System

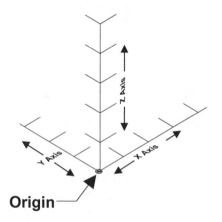

3D Axes Coordinate System

BASIC TYPES OF 3D MODELS

There are <u>3 basic types</u> of 3D models.

1. Wireframe models
2. Surface models
3. Solid models

A description of each starts on page 13-12.

---

*Note: Increased RAM requirement
The 3D visualization and rendering abilities of AutoCAD requires much more memory than ever before. It will run with 512MB but Autodesk recommends 2GB of RAM and a video card with at least 128MB of its own memory. Check Autodesk's website (www.autodesk.com) for the list of supported video cards.

ENTER AutoCAD 3D WORKSPACE

Note: AutoCAD LT has limited 3D capabilities and does not have a 3D workspace. LT users should skip to the projects starting after lesson 20.

1. Select the **3D Modeling Workspace.**
   A. Select the Workspace tool located in the lower right corner.

   B. Select the **3D Modeling** workspace.

2. Select **NEW** (to start a new file) and select one of the following templates from the list:
   acad3D.dwt (for Imperial units)
   Acadiso3D.dwt (for metric)

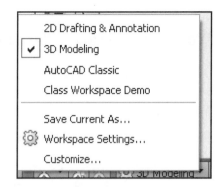

Now your screen should appear as the example shown below.

Notice the appearance of the Origin and the Cursor has changed. They now display the Z axis. Note: the arrows are always pointing in the positive direction

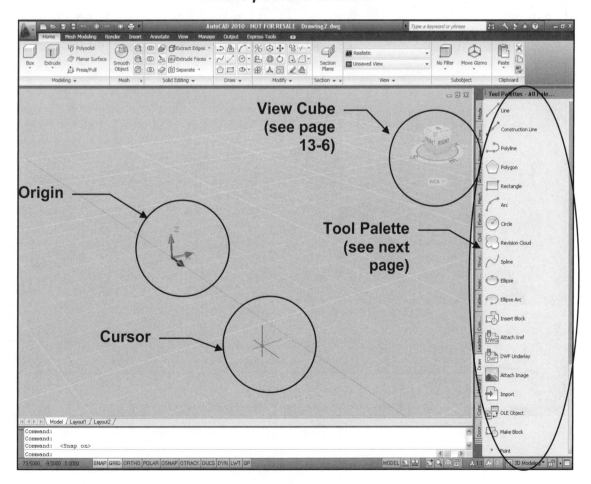

13-3

Tool Palette display

The Tool Palette displays many tools but I feel that it is redundant and unnecessary at this time. If you would prefer to expand the drawing area and not display the Tool Palette follow the instructions below. It is your choice.

You may easily toggle the Tool Palette ON or OFF by clicking on the Tool Palette tool.

Ribbon = View tab / Palettes panel /

Or

Type:
Toolpalettes <enter> for <u>ON</u>
Toolpalettesclose <enter> for <u>OFF</u>

Drawing area shown below **without** Tool Palette displayed.

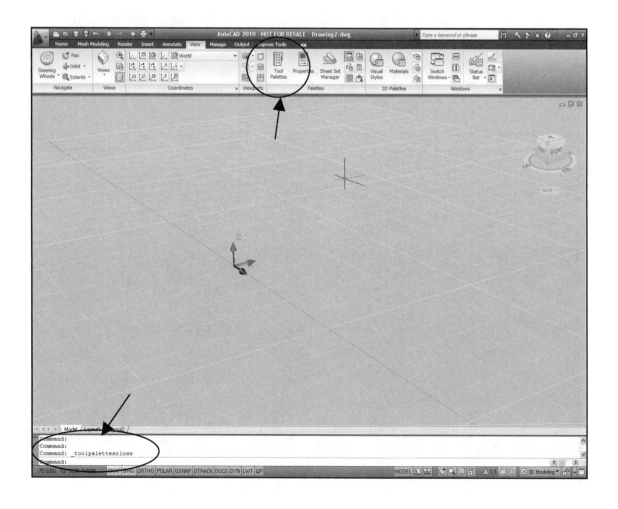

Viewing a 3D Model

Viewpoint

It is very important for you to be able to control how you view the 3D Model. The process of changing the view is called **selecting the Viewpoint**. (Note: View**point;** not view**port.**)
The **Viewpoint** is <u>the location where "**YOU**" are standing</u>. *This is a very important concept to understand*. <u>The Model does not actually move; you move around the model.</u> For example, if you want to look at the South East corner of your house, you need to walk to the South East corner of your lot and look at the corner of your house. Your house did not move, but you are seeing the South East corner of your house. If you want to see the Top or Plan View of your house, you would have to climb up on the roof and look down on the house. The house did not move, you did. So in other words, the Viewpoint is **your "Point of View"**.

Remember when you select a Viewpoint you are selecting where <u>you are standing</u>. The view that appears on the screen is what you would see when you move to the viewpoint and looked at the model.

How to change the Viewpoint
There are 3 methods to change the viewpoint of your model. **3D Orbit, 3D Views** and **ViewCube.** Each of the methods are described on the next pages.

ViewCube
Located in Drawing Area

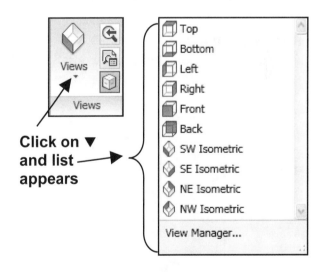

**Click on ▼
and list
appears**

3D Views
**Ribbon = View tab /
Views panel**

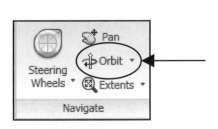

3D Orbit
**Ribbon = View tab /
Navigate panel**

ViewCube

The following is a description of the ViewCube tool and how to use it to view your model. But the best way to understand the ViewCube is to use it. So I would like you to **_OPEN_** the **_2010-3D Demo.dwg_** that you downloaded from our website. Refer to Intro-1 if you have not downloaded this file.

ViewCube Location
The ViewCube is located in the upper right corner of the drawing area. But you can change it's location. (Refer to ViewCube settings on page 13-8.)

Selecting a View
The ViewCube allows you to manipulate the model in order to see and work on all sides. I have added text to the model so you can see how clicking on a ViewCube, such as Top, automatically shows you the view of the top.

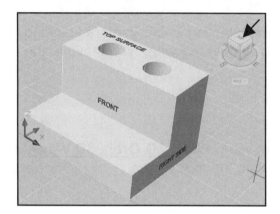
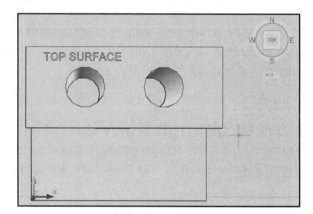

Rotating a View
After you have selected a view, such as Top, rotation arrows appear. If you click on either of the rotation arrows the view will rotate 90 degrees clockwise or counter clockwise.

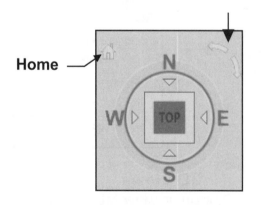
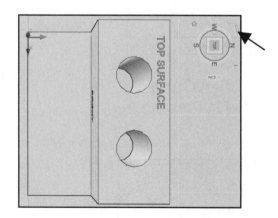

Returning to Original View
You may always return to the original **Home** view by selecting the little Home icon.
Note: The **Home** icon only appears when you place the cursor near the ViewCube.
You may also select a new view to become the **Home** view.
1. Orient the model to the preferred HOME appearance.
2. Right click on the ViewCube and select **Set current view as Home.**

ViewCube....continued

ViewCube Cursor Menu

If you right-click on the ViewCube the cursor menu will appear.

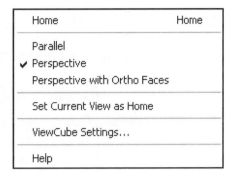

Home
This will return you to the original view that you specified as the Home View. This is the same as selecting the Home icon.

Parallel and Perspective
Displays the View as Parallel or Perspective.

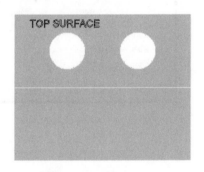 

Parallel   **Perspective**

Perspective with Ortho Faces
If the current view is a face view it will appear as parallel.

Set Current View as Home
Selects the current view as the Home view. When you select Home this view will be displayed.

ViewCube Setting...
Displays the ViewCube Settings dialog box shown on the next page.

Help
Takes you directly to the ViewCube content in the Help Menu

ViewCube....continued

ViewCube Settings

To change the ViewCube settings, and we will want to, <u>right click on the ViewCube</u> and select **ViewCube Settings...** from the list.

Most of these settings are personal preference. But I suggest that you uncheck the two boxes that I have shown. I think it will make the manipulation of the views smoother and less confusing. But again, it is your personal preference.

On-screen position:
Controls which corner the ViewCube is displayed.

ViewCube size:
Controls the size of the ViewCube display.

Inactive opacity:
Controls how visible the ViewCube is <u>when inactive only</u>. It will display dark automatically when you pass the cursor over it.

Show UCS menu:
Controls the display of the UCS drop-down menu below the ViewCube.

Snap to closest view:
When using Click and Drag to rotate the View you have a choice of it automatically snapping to the closest View or smoothly rotating to the desired View. *This is one that I prefer to uncheck in order to have a smooth rotation.*

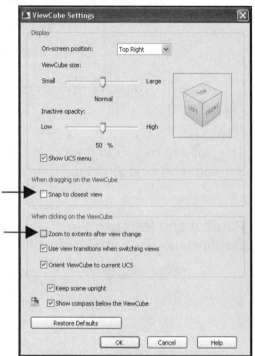

Zoom to extents after view change:
Each time you select a new View it also will Zoom to Extents if you select this option. *I prefer to control when Zoom is activated.*

Use View transitions when switching views:
Controls how smooth the transition appears when rotating.

Orient ViewCube to current UCS: (Important)
You definitely want this one ON. Everytime you move the Origin the Cube orients itself to match the new UCS. You will understand this one better in later lessons.

Keep scene upright:
Controls whether or not the view can be turned upside down. This reduces confusion.

Show compass below the ViewCube:
Controls the display of the compass, N, S, E, W.

> **Note: The settings stay with the computer not the drawing file.**

3D Orbit

The 3D Orbit allows you to rotate around the model using the click-drag method. This used to be the only way to freely rotate until ViewCube was added.

1. The 3D Orbit can be selected using one of the following:

    Ribbon = View tab / Navigate Panel /

    Keyboard = 3do

Select ▼

Description Below

2. Click and drag the cursor to rotate the model.

3. To stop press the **Esc** or **Enter** key

Constrained Orbit
Constrains 3D Orbit along the *XY* plane or the *Z* axis

Free Orbit
Orbits in any direction without reference to the planes.
The point of view is not constrained along the *XY* plane of the *Z* axis.

Continuous Orbit
Orbits continuously. Click and drag in the direction you want the continuous orbit to move, and then release the mouse button. The orbit continues to move in that direction.

> I add the constrained Orbit tool to my Quick Access Menu.
> Then I do not have to find it or type it.
> (Refer to Customizing the Quick Access tool bar Page 2-12.)

3D Views

Select **3D Views** as follows:

Ribbon = View tab / **Views** panel

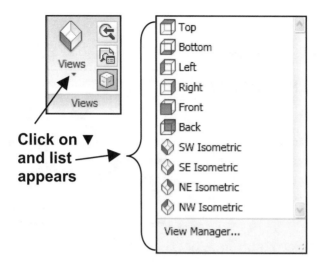

Click on ▼ and list appears

The list of view options are the same as the ViewCube.
If you have used AutoCAD's 3D software previously you are familiar with these.
Selecting from this list or using ViewCube is the users personal preference

Visual Styles

A Visual Style is a collection of settings that control the display of edges and shading.

1. Select the Visual Style Manager as follows:

    Ribbon = View tab / 3D Palettes panel / Visual Style

The **Visual Styles Manager** palette will appear on the left side of the drawing area. It is in the Auto-hide mode. If you pass your cursor over the bar the palette will be displayed.

The following pages will describe the Visual Styles and Face settings. The remainder of the selections will not be discussed but may be reviewed in the Help menu.

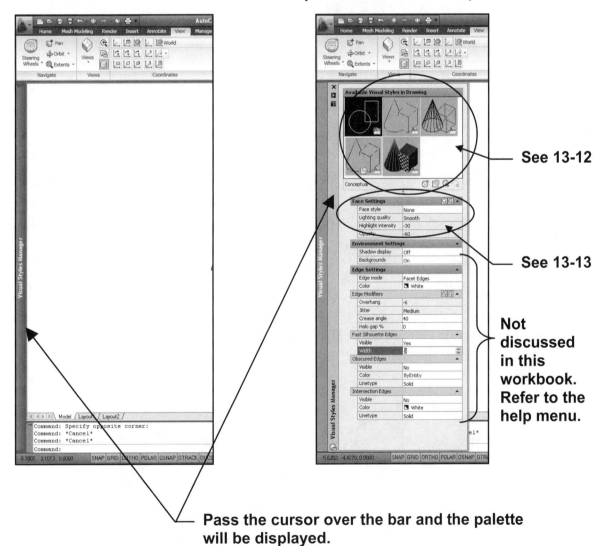

See 13-12

See 13-13

Not discussed in this workbook. Refer to the help menu.

**Pass the cursor over the bar and the palette will be displayed.
When you move the cursor off the palette it will hide again.**

Continued on the next page...

Visual Styles….continued

Description of Visual Styles

There are 5 visual styles supplied with the AutoCAD. Examples shown below.

> **2D Wireframe**: All lines and curves shown in Parallel
> **3D Hidden**: All lines and curves shown in Perspective. Hidden lines invisible.
> **3D Wireframe**: All lines and curves shown in Perspective.
> **Conceptual**: Shades the object with cool and warm using Gooch face style.
> **Realistic**: Shades the object with dark to light. May make details difficult to see.

Double click in any of the Visual Style boxes to select

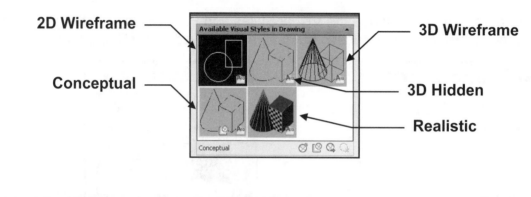

2D Wireframe — Available Visual Styles in Drawing — 3D Wireframe

Conceptual — 3D Hidden — Realistic

Conceptual

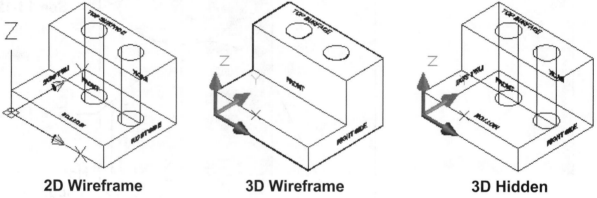

2D Wireframe   **3D Wireframe**   **3D Hidden**

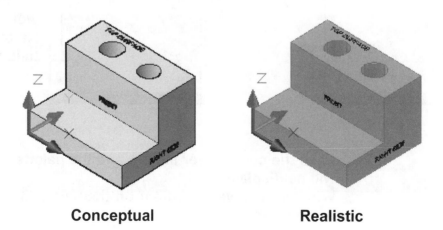

Conceptual   **Realistic**

Continued on the next page...

Visual Styles...continued

Face Settings
Controls the appearance of faces.

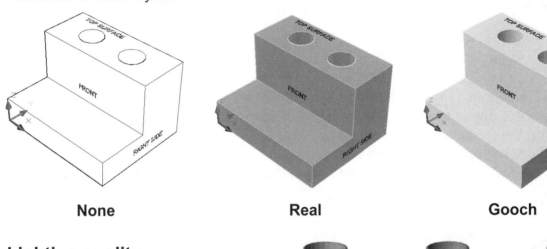

Face Settings	
Face style	None
Lighting quality	Smooth
Highlight intensity	-30
Opacity	-60

When you select a Visual Style the Face settings will change also. Select each of the Visual Styles shown on the previous page and notice the Face settings.

Face Style
There are 3 Face Styles.

None	**Real**	**Gooch**

Lighting quality
There are 3 Lighting Qualities.
Faceted: faceted appearance.
Smooth: regular quality smooth appearance. (This is the default level)
Smoothest: High quality smooth appearance.

faceted	smooth	smoothest

Highlight Intensity
The size of the highlights on an object affects the perception of shininess. A smaller, more intense highlight makes objects look shinier.

Highlight Intensity: off	size: 10	size: 30

Opacity
The opacity property controls the transparency of an object.

Opacity: off	Opacity: 20

WIREFRAME MODEL

A wireframe model of a box is basically 12 pieces of wire (lines). Each wire represents an **edge** of the object. You can see through the object because there are no surfaces to obscure your view. This type of model does not aid in the visualization of the 3D object. Wireframe models have no volume or mass.

How to draw a Wireframe Box

1. Start a **NEW** file using **acad3D.dwt**
2. Draw a **Rectangle** just like you would in 2D.   **First corner: 0,0,0 Second corner: 6,4**
3. Zoom in closer if the rectangle seems too small.

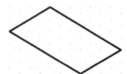

4. Copy the Rectangle 5" **above** the original rectangle as follows:
   a. Select the **Copy** command.
   b. Click on the Rectangle then <enter>.
   c. Select the **basepoint** as shown below.
   d. Type the **X, Y, Z** coordinates for the new location: @ **0, 0, 5 <enter> <enter>**

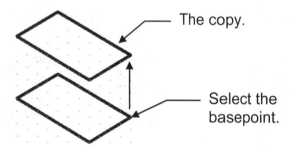

The copy.

Select the basepoint.

*Now think about the coordinates entered. The coordinates **0, 0, 5** mean that you **do not** want the new rectangle location to move in the **X** or **Y** axis. But you **do** want the new rectangle location to move **5"** in the **positive "Z"** axis.*

5. Now using **Line** command, draw lines from the corners of the lower rectangle to the upper rectangle.

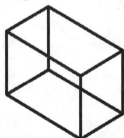

This wireframe is displayed as "Parallel".

That is all there is to it. You have now completed a Wireframe Box.
This Wireframe can now be used as the structure for a Surface Model described on the next page.

SURFACE MODELS

A **Surface Model** is like an <u>empty cardboard box</u>. All surfaces and edges are defined but there is nothing inside. The model appears to be solid but it is actually an empty shell. The hidden line removal command can be used because the front surfaces obscure the back surfaces from view. A surface model makes a good visual <u>representation</u> of a 3D object.

You may attach a 3D Surface to a Wireframe structure or you may use one of AutoCAD's pre-defined 3D Objects.

How to add a 3D Surface to a Wireframe structure

1. Create a Wireframe Model as shown on the previous page.

2. Type: **3dface <enter>**

3. Draw a 3D Surface by snapping to the 4 corners as shown below and **<enter>.**
   (The Shape will **Close** automatically.)
   Note: Each surface must be created as a separate object.

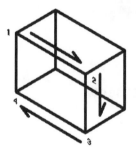 

Your 3D Wireframe model now has 2 surfaces attached.
Now you may even use a "Visual Style".

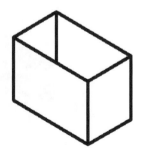 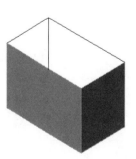

SOLID MODELS *(Refer to Lesson 14 for detailed instructions)*

*It is important that you understand both **Wireframe** and **Surface** Modeling **but** you will not be using either very often. **Solid Modeling** is the most useful and <u>the most fun</u>.*

Solid models have edges, surfaces and **mass**.

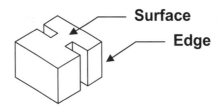

Solid Models can be modified using many **Solid Editing** features such as the ones described below.

Use Boolean operations such as **Union, Subtract and Intersect** to create a solid form.

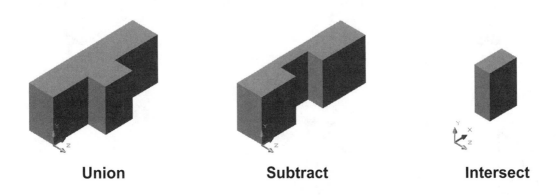

| **Union** | **Subtract** | **Intersect** |

| Shapes can be: | **Revolved** | **Shelled** |

***You will learn all of these features and more in the following Lessons.
But first do the following Exercises to get some practice with
Wireframe and Surface Modeling.***

EXERCISE 13A

INSTRUCTIONS:

1. Start a **NEW** file using **acad3D.dwt.**

2. Create the **Wireframe** Model shown below. (Refer to page 13-14)

3. Display the Model as follows:

> **SE Isometric** (Refer to page 13-10)
>
> Or
>
> click on the corner of the ViewCube ———

Note:
The rectangle may appear small. Zoom in.
Also, change the color if the default color is
too light. Use "Properties".

4. Display the Model as **Parallel** not perspective. (Refer to page 13-7)

5. Do not dimension.

6. **Save** the drawing as: **EX13A**

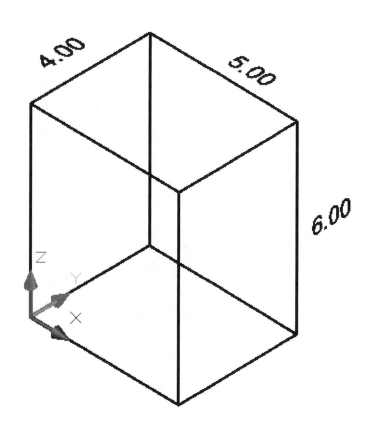

EXERCISE 13B

INSTRUCTIONS:

1. Open **EX13A**

2. Add surfaces to all 4 sides and Top and Bottom. (Refer to page 13-15)

   (Use ViewCube or 3d Orbit to rotate the model to access each side.)

3. Display as

   a. **SE Isometric**

   b. Visual Style **Conceptual** (Refer to page 13-12)

   c. **Parallel**.

4. **Save** the drawing as: **EX13B**

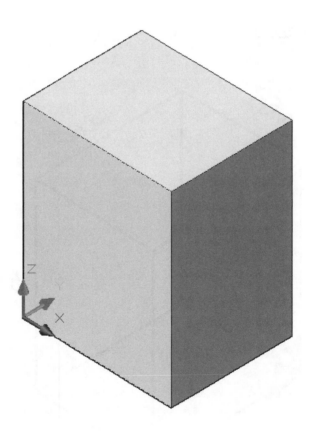

LEARNING OBJECTIVES

After completing this lesson, you will be able to:

Construct 7 Solid model Primitives:
   Box, Sphere, Cylinder, Cone, Wedge, Torus and Pyramid

Sorry LT users, you do not have this option.

When you have completed this lesson you will be able to create the 3D Model shown below

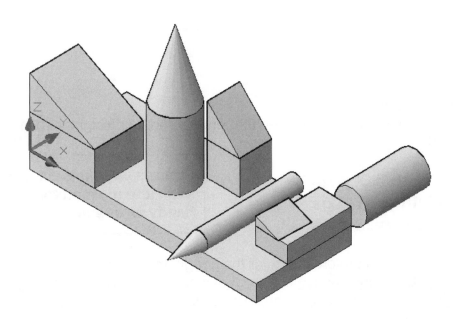

LESSON 14

DRAWING BASIC GEOMETRIC SHAPES

In this lesson you will learn the required steps to construct each of AutoCAD's basic geometric shapes. Each one requires different input information and some have multiple methods of construction.

AutoCAD has <u>7 Solid shapes</u>.
<u>Cylinder, Cone, Sphere, Box, Pyramid, Wedge and Torus</u>.

To select a solid primitive use one of the following:

> **Ribbon = Home tab / Modeling panel / ▼**

> **Keyboard = Type the name such as: Box**

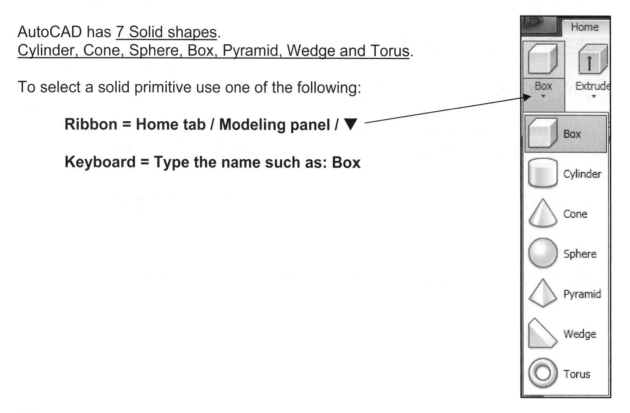

3D input direction
When drawing in 3D and prompted for Length, Width or Height, each input corresponds to an axis direction as follows:

Length = X Axis
Width = Y Axis
Height = Z Axis

> **I always write this on a little post-it and stick it to my monitor. It comes in handy as a reminder.**

For example, if you are prompted for the <u>Length</u> the dimension that you enter will be drawn on the <u>X axis</u>. <u>So keep your eye on the UCS icon in order to draw the objects in the correct orientation.</u> Although it should be easy to visualize because you will actually see the shape constructed as you draw.

Consider starting the primitives on the Origin. It is useful to know where the primitive is located so it can be moved or rotated easily.

In Lesson 15 you will learn more about moving the UCS around to fit your construction needs. But just relax and let's take it one step at a time.

BOX

There are 4 methods to draw a **Solid Box**. Which one you will use will depend on what information you know. For example, if you know where the corners of the base are located and the height, then you could use method 1 or 2.

Start a new file and select Acad3d.dwt

Method 1 (You will enter the location for: base corner, diagonal corner and height)
a. Select the **SE Isometric** view.
b. Select the **Box** command. (See page 14-2)
c. Specify first corner [Center] <0,0,0>: *type coordinates or pick location with cursor (P1).*
d. Specify other corner or [Cube/Length]: *type coordinates for the diagonal corner or pick location with the cursor (P2).*
e. Specify height or [2Point]: *type or use cursor.*

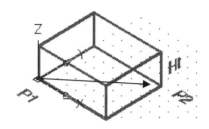

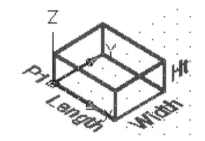

Method 2 (You will enter the dimension for L, W, and Ht)
a. Select the **SE Isometric** view.
b. Select the **Box** command.
c. Specify first corner or [Center] <0,0,0>: *type coordinates or pick location with cursor. (P1)*
d. Specify other corner or [Cube/Length]: *type "L" <enter>.*
e. Specify length: *enter the Length (X axis).*
f. Specify width: *enter the Width (Y axis).*
g. Specify height: *enter the Height (Z axis).*

(Note: when entering L, W or H use DDE with Ortho "ON".)

Method 3 (If Length, Width & Height have the same dimension)
a. Select the **SE Isometric** view.
b. Select the **Box** command.
c. Specify first corner or [Center] <0,0,0>: *type coordinates or pick location with cursor. (P1)*
d. Specify corner or [Cube/Length]: *type "C" <enter>.*
e. Specify length: *enter the dimension.*

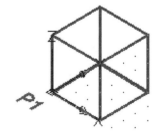

Method 4 (You will enter the location for the center, a corner and the height)
a. Select the **SE Isometric** view.
b. Select the **Box** command.
c. Specify corner of box or [Center] <0,0,0>: *type "C"<enter>.*
d. Specify center of box <0,0,0>: *type coordinates or pick location with cursor. (P1)*
e. Specify corner or [Cube/Length]: *type coordinates for a corner or pick location with the cursor. (P2)*
f. Specify height: *type the height.*
(Note: The total Length, Width and Height straddles the centerpoint.)

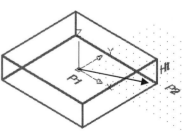

SPHERE

Sphere creates a spherical solid. You define the center point and then define the size by entering either the radius or the diameter.

a. **Start a new file and select <u>Acad3d.dwt</u>**.
b. Select the **Sphere** command. (See page 14-2)
c. Specify center point or [3P/2P/Ttr]: *type coordinates or pick location with cursor.*
d. Specify radius of sphere or [Diameter]: *enter radius or D.*

You may enjoy experimenting with Visual Styles.

CYLINDER

Cylinder creates a cylindrical solid. You will define the center location for the base, define the radius or diameter and then the height or location for the other end.
Start a new file and select <u>Acad3d.dwt</u>

The two methods below are the most commonly used.

Method 1
The default orientation of the cylinder has the base on the X,Y plane and the height is in the Z direction. When you enter the height, the cylinder grows in the Z axis direction. You may enter a positive or negative number. It depends upon which direction you want the cylinder to grow. The Z-axis arrow points to the positive direction.
So remember, keep an eye on the 3D UCS icon.

a. Select the **SE Isometric** view.
b. Select the **Cylinder** command.
c. Specify center point for base or [3P/2P/Ttr/Elliptical] <0,0,0>: *type coordinates or pick*
location with cursor (P1)
d. Specify radius for base of cylinder or [Diameter]: *enter radius or D. (P2)*
e. Specify height of cylinder or [Center of other end]: *enter the height (Ht). (P3)*

Remember a "positive" input is in the direction the arrow points.
A negative input is the opposite.
The example to the left has a positive Height input.

Method 2
The orientation of the Cylinder base depends on the placement of the <u>Center of the other End</u>. This method allows you to tilt the cylinder. Define the center of the base and radius then select the "Axis Endpoint" option. Define the "Axis Endpoint" entering relative coordinates or snapping to an object. (Soon you will learn how to rotate the Origin)

a. Select the **SE Isometric** view.
b. Select the **Cylinder** command.
c. Specify center point of base or [3P/2P/Ttr/Elliptical] <0,0,0>: *type coordinates or pick*
location with cursor (P1).
d. Specify base radius or [Diameter]: *enter radius or D. (P2)*
e. Specify height or [2 point/Axis endpoint]: *A <enter>.*
f. Specify axis endpoint: *type coordinates or snap to an object. (P3)*
(See the 2 examples below)

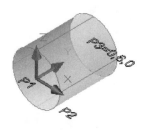

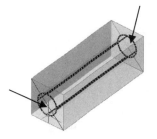

<u>Enter coordinates (0,6,0</u>   <u>Snap to an object (Lines)</u>

CONE

Cone creates a Conical solid. There are 2 methods to create a Cone.
You will define the center location, radius or Diameter for the base and then define the height or location for the apex. **Start a new file and select <u>Acad3d.dwt</u>**

Method 1
The default orientation for the base is on the X and Y plane and the height is perpendicular in the Z direction.

a. Select the **SE Isometric** view.
b. Select the **Cone** command. (See page 14-2)
c. Specify center point of base [3P/2P/Ttr/Elliptical] <0,0,0>: *type coordinates or pick location with cursor (P1)*
d. Specify base radius or [Diameter]: *enter radius or D. (P2)*
e. Specify height or [2Point/Axis endpoint/Top Radius]: *enter the height (P3) (can be positive or negative).*

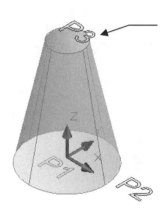

If you select "Top Radius" you may enter a radius for the top and then the height.

Method 2
The orientation of the Cone depends on the placement of the Apex. Define the center of the base and radius then select the "Apex" option. Define the "Apex" location using coordinates or snapping to an object.

a. Select the **SE Isometric** view.
b. Select the **Cone** command.
c. Specify center point of base or [Elliptical] <0,0,0>: *type coordinates or pick location with cursor (P1)*
d. Specify base radius or [Diameter]: *enter radius or D. (P2)*
e. Specify height or [2Point/Axis endpoint/Top radius]: *type "A" <enter>.*
f. Specify axis endpoint: *type coordinates or snap to an object. (P3)*

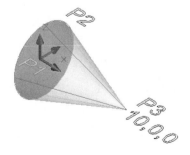
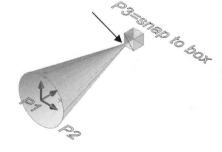

<u>**Enter coordinates**</u>   <u>**Snap to an object**</u>

WEDGE

Wedge creates a wedge solid. There are 4 methods to create a Wedge.
The base is always parallel with the current UCS, XY plane, and the height is always along the Z axis.   The slope is always from the Z axis along the X axis.
Start a new file and select <u>Acad3d.dwt</u>

Method 1 (Define the location for 2 corners of the Base and then the Height)
a. Select the **SE Isometric** view.
b. Select the **Wedge** command. (See page 14-2)
c. Specify first corner or [Center] <0,0,0>: *type coordinates or pick location with cursor. (P1)*
d. Specify other corner or [Cube/Length]: *type coordinates for the diagonal corner or pick location with the cursor. (P2)*
e. Specify height: *type the height*

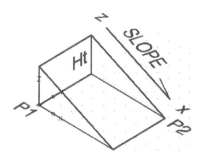

Method 2 (Define the location for each: L, W and Ht)
a. Select the **SE Isometric** view.
b. Select the **Wedge** command.
c. Specify first corner or [Center] <0,0,0>: *type coordinates or pick location with cursor. (P1)*
d. Specify other corner or [Cube/**Length**]: *type "L" <enter>.*
e. Specify length: *enter the Length (X axis).*
f. Specify width: *enter the Width (Y axis).*
g. Specify height: *enter the Height (Z axis).*

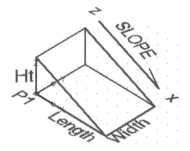

Method 3 (Define the same dimension for Length, Width & Height)
a. Select the **SE Isometric** view.
b. Select the **Wedge** command.
c. Specify first corner or [Center] <0,0,0>: *type coordinates or pick location with cursor. (P1)*
d. Specify other corner or [**Cube**/Length]: *type "C" <enter.>*
e. Specify length: *enter the dimension.*

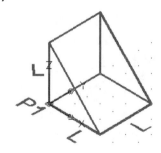

Method 4 (Define the location for the Center and the Height)
a. Select the **SE Isometric** view.
b. Select the **Wedge** command.
c. Specify first corner or [Center] <0,0,0>: *type "C"<enter>.*
d. Specify center <0,0,0>: *type coordinates or pick location with cursor. (P1)*
e. Specify corner or [Cube/Length]: *type coordinates for a corner or pick location with the cursor. (P2)*
f. Specify height: *type the height.*

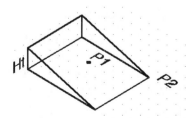

TORUS

Torus can be used to create 3 different solid shapes.

Start a new file and select <u>Acad3d.dwt</u>

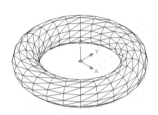

Torus = 3 Tube = 1
Donut shaped

Torus = -2 Tube = 3
Football shaped

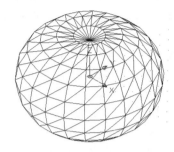

Torus = 1 Tube = 3
Self-Intersecting

The 2 dimensions that are required are the radius
or diameter of the **Torus** and the **Tube**.
(Pay close attention to what the Torus and Tube are
defining. Example on the right)

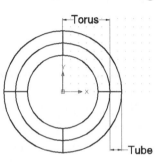

Donut shaped
Note: the <u>Torus radius</u> must be <u>greater than</u> the <u>Tube radius</u>.
a. Select the **SE Isometric** view.
b. Select the **Torus** command. (See page 14-2)
c. Specify center point or [3P/2P/Ttr]: *type coordinates or pick location with cursor.*
d. Specify radius or [Diameter]: *(this dim. must be <u>greater than</u> the <u>Tube</u> radius).*
e. Specify tube radius or [2pt/Diameter]: *(this dim. must be <u>less than</u> the <u>Torus</u> radius).*

Football shaped
Note: the <u>Torus radius</u> must be <u>negative</u> and the <u>Tube radius positive</u> and <u>greater than</u> the <u>Torus radius</u>.
a. Select the **SE Isometric** view.
b. Select the **Torus** command.
c. Specify center point or [3P/2P/Ttr]: *type coordinates or pick location with cursor.*
d. Specify radius or [Diameter]: *(this dim. must be <u>negative</u>).*
e. Specify tube radius or [2pt/Diameter]: *(this dim. must be <u>positive</u> and <u>greater than the Torus radius</u>).*

Self-Intersecting
Note: the <u>Torus radius</u> must be <u>less than</u> the <u>Tube radius</u>.
a. Select the **SE Isometric** view.
b. Select the **Torus** command.
c. Specify center point or [3P/2P/Ttr]: *type coordinates or pick location with cursor.*
d. Specify radius or [Diameter]: *(this dim. must be <u>less than</u> the <u>Tube</u> radius).*
e. Specify tube radius or [2pt/Diameter]: *(this dim. must be <u>greater than</u> the <u>Torus</u> radius).*

PYRAMID

The method for drawing a Pyramid is very similar to drawing a polygon for the base and a cone for the height. You specify the number of sides the base will need. It may have from 3 to 32 sides. Decide if the base radius is Inscribed or Circumscribed. Then define the height. If you select the Top option you can truncate it. **Start a new file and select <u>Acad3d.dwt</u>**

a. Select **SE Isometric** view.
b. Select the **Pyramid** command.
c. Specify center point of base or [Edge/Sides]: *type "S"<enter>*
d. Enter number of sides <4>: *type number of sides <enter>*
e. Specify center point of base or [Edge/Sides]: *type coordinates or pick location with cursor for Center location*
f. Specify base radius or [Inscribed] <1.000>: *enter radius or D*
g. Specify height or [2Point/Axis endpoint/Top radius] <1.000>: *enter the height (can be positive or negative).*

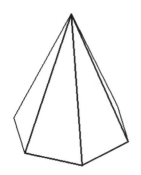

Pyramid with 6 sided base

Truncated Pyramid
Use the "Top radius" option.
Top radius is smaller than base radius.

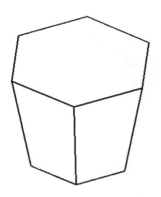

Truncated Pyramid
Use the "Top radius" option.
Top radius is larger than base radius.

EXERCISE 14A
Create 4 Solid Boxes

1. **Select File / New and select <u>Acad3d.dwt</u>**

2. Select the **SE Isometric view**

3. **Create 4 solid boxes as shown below. *You decide which method to use.*
   (Refer to page 14-3 for instructions if necessary)**

4. Save as **EX-14A**

5. **Do not Dimension** *Do not add the letters B, C or D. They are for reference only.*

BOX A
L = 14 W = 6 HT = -1

BOX B
L = 4 W = 3 HT = 2

BOX C
ALL SIDES 2"

BOX D
L = 2 W = 4 HT = 1 (Think, positive or negative)

*Enter X, Y, Z coordinates for <u>each point</u> or Dynamic Input. If you use Dynamic Input
make sure <u>Ortho</u> is ON for <u>each point</u>.*
REMEMBER   L = X AXIS   W = Y AXIS   HT = Z AXIS

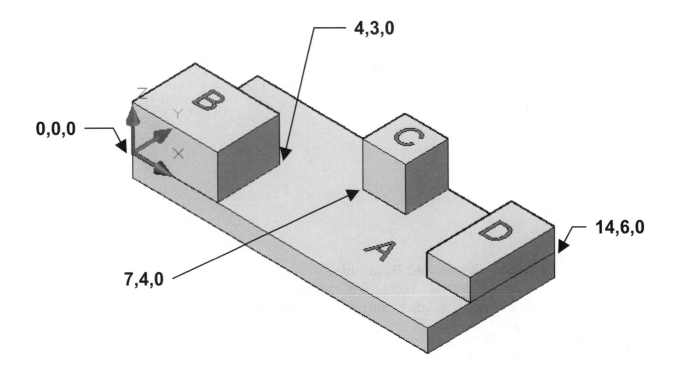

EXERCISE 14B
Create 3 solid Cylinders

1. Open **EX-14A**

2. Add the 3 Cylinders as shown below.
   Use Method 1 or 2 shown on page 14-5.
   The method will depend on the information you are given below.

   Read the command line prompts to make sure you are entering the correct information.

3. Select 3D Orbit to confirm you have placed the objects in the correct location.

4. Save as **EX-14B**

   Do not Dimension

CYLINDER E	CYLINDER G	CYLINDER H
Radius = 1.25	6" from end to end	Radius = 1
Ht = 4"	Dia = 1 (notice "diameter")	Length = 4
	Hint: Use Axis endpoint	
	@0,6,0	

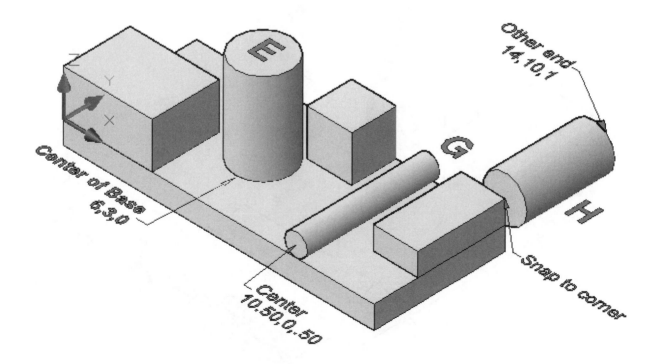

EXERCISE 14C
Create 2 solid Cones

1. Open **EX-14B**

2. Add the 2 Cones as shown below.
   (Refer to page 14-6 for instructions if necessary)

3. Save as **EX-14C**

   Do not Dimension

CONE J	**CONE K**
Radius = 1.25	Radius = .50
Ht = 4"	Ht = 2"

Note:
Locate the Center location by snapping to the Center of the Cylinder.
Locate the Radius by snapping to the Quadrant of the Cylinder.
Remember, Ortho ON will help.

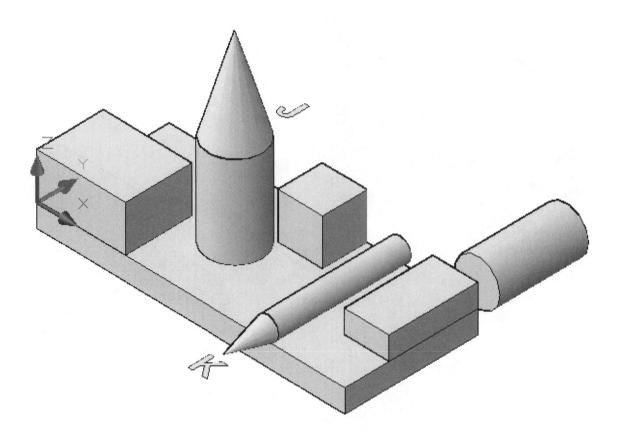

EXERCISE 14D
Create 3 solid Wedges

1. Open **EX-14C**

2. Add the 3 Wedges as shown below.
   (Refer to page 14-7 for instructions if necessary)

3. Save as **EX-14D**

   Do not Dimension

WEDGE L	**WEDGE M**	**WEDGE N**
Ht = 2"	L, W & H = 2	L = 1.5
		W = 2
		Ht = .5

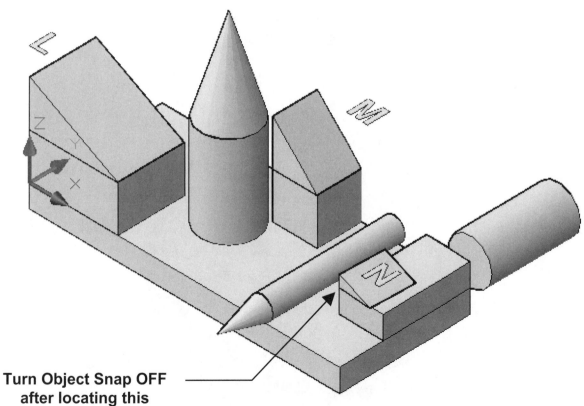

**Turn Object Snap OFF
after locating this
corner. It may interfere
with placing the other
corner.**

EXERCISE 14E
Create a solid Sphere

1. **Select File / New and select <u>Acad3d.dwt</u>**

2. Select the **SE Isometric view**

3. **Create the solid Sphere shown below.**
   (Refer to page 14-4 for instructions if necessary)

4. Experiment with Visual Styles.

5. Save as **EX-14E**

   Do not Dimension

Center = 0, 0, 0
Radius = 4

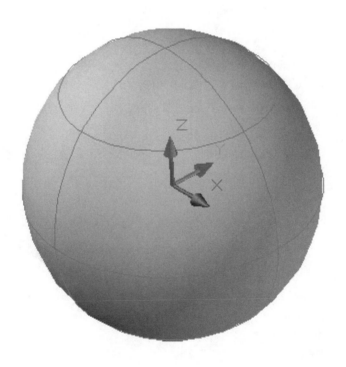

EXERCISE 14F
Create 3 solid Torus'

1. **Select File / New and select <u>Acad3d.dwt</u>**

2. Select the SE Isometric View.

3. Add the 3 Torus' as shown below.
   (Refer to page 14-8 for instructions if necessary)

4. Save as **EX-14F**   **Do not Dimension**

DONUT	**FOOTBALL**	**SELF-INTERSECTING**
Center = 4, 3, 0	Center = 8.5, 7, 0	Center = 10.5, 3, 0
Torus Rad = 3	Torus Rad = -3	Torus Rad = 1
Tube Rad = 1	Tube Rad = 5	Tube Rad = 1.50

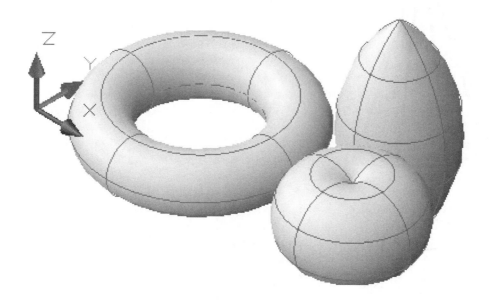

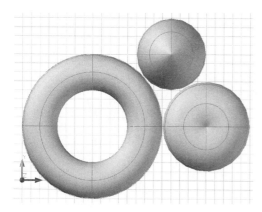

Take a look at the Top View to see if you have them positioned correctly.
1. Select **<u>Parallel</u>** instead of **<u>Perspective</u>**
   (Refer to page 13-7)

EXERCISE 14G
Create 2 Pyramids

1. **Select File / New and select <u>Acad3d.dwt</u>**

2. Draw the 2 Pyramids (INSCRIBED) shown below.
   (Refer to page 14-9 for instructions if necessary)

3. Save as **EX-14G**

   Do not Dimension

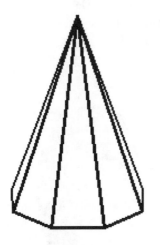

Pyramid 1
Sides = 8
Base Radius = 2
Ht = 6

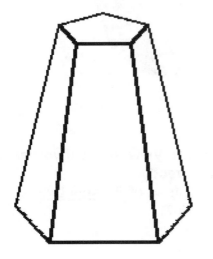

Pyramid 2
Sides = 5
Base Radius = 2
Top Radius = 1
Ht = 6

LEARNING OBJECTIVES

After completing this lesson, you will be able to:

1. Move the UCS to aid in the construction of the model.
2. Understand and use Boolean operations:
    Union, Subtract and Intersect.

Sorry LT users, you do not have this option.

LESSON 15

CONFIGURING OPTIONS FOR 3D

In the Intro section of this workbook you configured your system for use with this workbook. But I skipped the 3D section. Now you need to make those selections.

1. Type **Options <enter>**

2. Select the **3D Modeling** tab.

3. Change the settings to match the settings shown below.

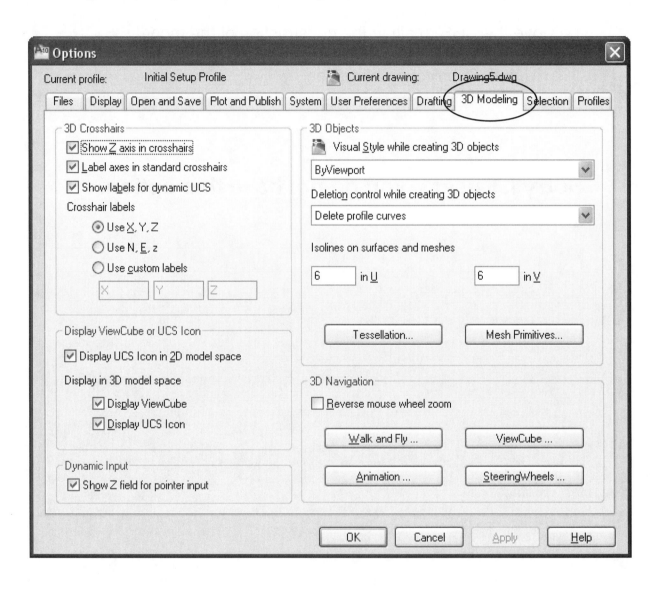

UNDERSTANDING THE UCS

In this Lesson you are going to learn how to manipulate the <u>UCS Origin</u> to make constructing 3D models easy and accurate.

World Coordinate System vs. User Coordinate System

All objects in a drawing are defined with XYZ coordinates measured from the 0,0,0 Origin. This coordinate system is **fixed** and is referred to as the **World Coordinate System (WCS)**. When you first launch AutoCAD, the WCS icon is in the lower left corner of the screen. (An easy way to return to WCS is: Type UCS <enter> <enter>.)

The **User Coordinate Sytem (UCS)** allows you to move a temporary Origin to any location. (This procedure is referred to as "moving the Origin".)

Why move the UCS Origin?

<u>Objects that are drawn will always be parallel to the XY plane</u>. (Unless you type a Z coordinate) So it is necessary to define which plane is the XY plane. You define the plane by moving the UCS origin to a surface.

Let me give you a few examples and maybe it will become more clear.

Example 1:

This wedge was drawn using L, W, H. When you entered the L, W and H how did AutoCAD know to draw its base parallel to the XY plane? Because <u>all objects are drawn parallel to the XY plane</u>. Length is always in the X axis, Width is always in the Y axis and Height is in the Z axis. Notice the UCS icon displays the plane orientation.

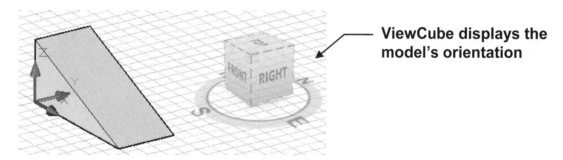

ViewCube displays the model's orientation

Example 2:

Below I have drawn the base of the cylinder parallel to the sloped surface. How did I do that? Notice the UCS icon. I moved it to the surface I wanted to draw on. (Remember: all objects are drawn parallel to the XY plane.) The base of the cylinder is automatically placed on the XY surface I defined and the height is always drawn in the Z axis.

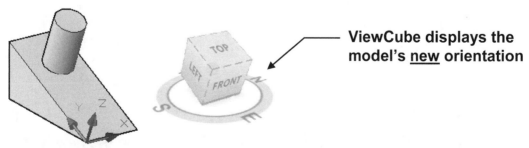

ViewCube displays the model's <u>new</u> orientation

How did I move the UCS icon to the orientation shown in the example above? I will discuss that next.

MOVING the UCS ICON

There are many options available to manipulate the UCS icon. In this Lesson the two most commonly used options will be discussed. One is to move the UCS permanently and one is to move the UCS temporarily.

HOW TO MOVE THE UCS ICON PERMANENTLY USING "3 POINT".

Use this method if you need to add a feature to a surface "in a specific location".

In the example below the UCS is moved to the corner of the sloped surface. This will allow you to place a feature, such as the cylinder, using X, Y coordinates from the Origin.

1. Select the **3 point** command using one of the following:

    Ribbon = View tab / Coordinates panel /

    Keyboard = UCS <enter> 3 <enter>

2. Specify new origin point <0,0,0>: *Locate where you want the new 0,0,0 (P1- place with the cursor using object snap or type coordinates---be accurate)*

3. Specify point on positive portion of X-axis <1.000,0.000,0.000>: *define which direction is positive X axis by snapping to a corner (P2).*

4. Specify point on positive-Y portion of the UCS XY plane <0.000,1.000,0.000>: *define which direction is positive Y axis by snapping to a corner (P3)*

<u>**2 EXAMPLES**</u>

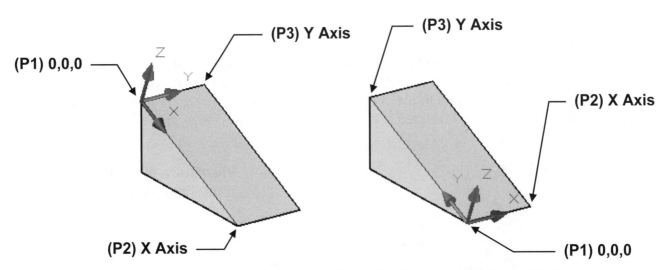

On your screen, the Grid should be displayed on the new X, Y plane.
Use 3d Orbit to get a good look.

MOVING THE UCS ICON TEMPORARILY USING "DYNAMIC UCS"

Use this method if you are placing a feature on a surface but you are not concerned with location at this time. You may move the feature later.
In the previous you must return the UCS Origin to its original location. This method allows you to <u>dynamically</u> change the XY Plane temporarily and then it is returned to the previous location automatically.

1. You may turn this feature **ON** or **OFF** with the **DUCS** 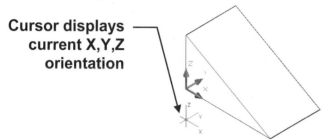 button on the status bar or **F6** key.

Cursor displays current X,Y,Z orientation

2. <u>Start a command</u> such as Cylinder. (You **must** be in a command)

3. <u>Pass the cursor over a face</u>. (Such as the slope of a wedge)
    A faint <u>dashed border</u> appears around the selected face and the cursor displays the new X,Y and Z planes.

Cursor displays selected surface X,Y,Z planes.

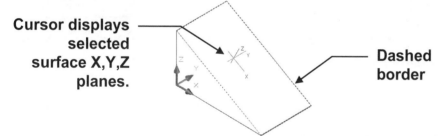

Dashed border

4. <u>Continue with the command</u>. (Such as click to place the center of the cylinder) The UCS icon temporarily moves to that location and temporarily orients the X,Y plane.

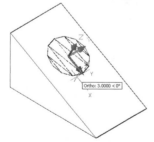

5. <u>Complete the command</u>. (Such as specify the length) The UCS icon automatically returns to its previous location and the cursor displays the current X,Y,Z orientation.

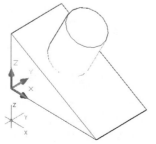

ROTATING the UCS ICON

Rotating the UCS is another option to manipulate the UCS icon. You may rotate around any one of the 3 axes. You may use the cursor to define the rotation angle or type the rotation angle.

1. Select the UCS Rotate command using one of the following:

    Ribbon = View tab / Coordinates panel /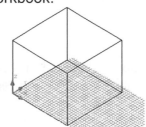

    Keyboard = UCS <enter> X, Y or Z <enter>

2. Select which Axis to rotate "**Around**".
    *(For Example: If you select **X**, you are rotating the Y and Z-axes around the X axis.)*

3. Specify rotation angle about X axis <90>: ***type the rotation angle or use cursor.***

Understanding the rotation angle

The easiest way to determine the rotation angle is to think of you standing in front of the point of the axis arrow, and then rotate the other 2 axes.
For example, if you are rotating around the X-axis put the point of the X-axis arrow against your chest and then grab the Z axis with your left hand and the Y axis with your right hand. Now rotate your hands Clockwise or Counterclockwise like you are holding a steering wheel vertically. The X-axis arrow, corkscrewing into your chest, is the Axis you are rotating around. **Clockwise is negative** and **Counter Clockwise is positive** just like the Polar clock you learned about in the Beginning workbook.

Original UCS Position

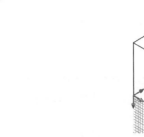

Positive Z direction

Positive Z direction

CCW Rotation 90

CW Rotation -90

Notice the grid indicates the drawing plane.

Note: To return the UCS to its default location and rotation: type UCS <enter> <enter>.
    If its stubborn select the SE Isometric View and then type UCS <enter> <enter>

NEW DIRECTION FOR Z AXIS

Remember that "**Height**" is always the **positive** Z axis. Sometimes the positive Z axis is not oriented to suit your height direction need. The "**Zaxis Vector**" option allows you to change the positive direction of the Z axis. The X and Y axes will change also automatically. (Turn OFF the **DUCS** function. It will be confusing to use both. See 15-5)

1. Select the **Zaxis** command using one of the following:

> **Ribbon = View tab / Coordinates panel /**

> **Keyboard = UCS / ZA**

2. Specify new origin point or object <0,0,0>: *Place the new Origin location*

3. Specify point on positive portion of Z-axis <0.000,0.000,1.000>: *select the positive Z direction.*

EXAMPLES
Notice all 3 examples below use the same information to draw the Cylinder. The only difference is the direction of the Z-axis.

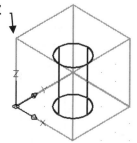

The Original
The XY drawing plane is on the <u>inside bottom</u>.
The Cylinder is drawn using the following:
Base center location: **2, 2, 0**
Radius : **1**
Ht = **3**

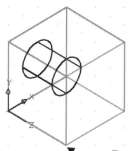

Positive Zaxis direction has been changed
The XY drawing plane is on the <u>inside of the left side</u>.
The Cylinder is drawn using the following:
Base center location: **2, 2, 0 (same as above)**
Radius : **1 (same as above)**
Ht = **3 (same as above)**

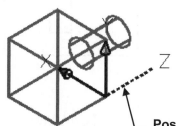

Positive Zaxis direction has been changed again
The XY drawing plane is on the <u>outside of the Back</u>.
The Cylinder is drawn using the following:
Base center location: **2, 2, 0 (same as above**
Radius : **1 (same as above)**
Ht = **3 (same as above)**

BOOLEAN OPERATIONS   (Sorry, not available in version LT)

Solid objects can be combined using Boolean operations to create **composite solids**.
AutoCAD's **Boolean** operations are **Union, Subtract** and **Intersect.**

> **Boolean** is a math term which means using logical functions, such as addition or subtraction on objects.
> George Boole (1815-1864) was an English mathematician. He developed a system of mathematical logic
> where all variables have the value of either one or zero. Boole's two-value logic, or binary algebra, is the
> basis for the mathematical calculations used by computers and is required in the construction of
> composite solids.

UNION
The **Union** command creates one solid object from 2 or more solid objects.

1. Select the **Union** command using one of the following:

   Ribbon = Home tab / Solid Editing /

   Keyboard = uni

2. Select objects: *select the solids to be combined <enter>*

BEFORE UNION – 5 separate objects

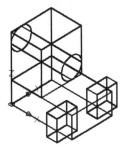
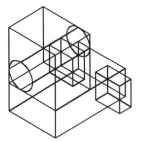
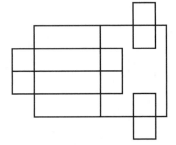

| **SE Isometric** | **SW Isometric** | **Top View** |

AFTER UNION – 1 object

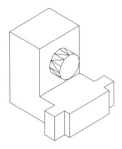
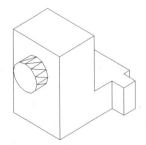
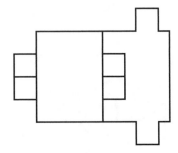

| **SE Isometric** | **SW Isometric** | **Top View** |

SUBTRACT

The **Subtract** command subtracts one solid from another solid.

1. Select the **Subtract** command using one of the following:

   Ribbon = Home tab / Solid Editing /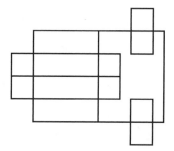

   Keyboard = su

2. Select solids and regions to subtract from…
   Select objects: ***select the solid object to subtract <u>from</u>***
   Select objects: ***<enter>***

3. Select solids and regions to subtract…
   Select objects: ***select the solid object to subtract***
   Select objects: ***<enter>***

BEFORE SUBTRACT – 5 separate objects

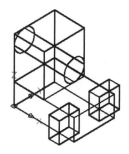 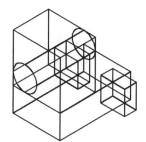 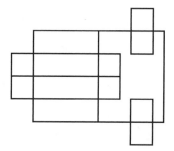

| SE Isometric | SW Isometric | Top View |

AFTER SUBTRACT – 1 object

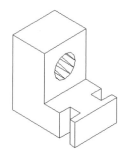 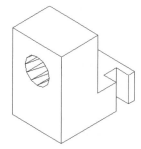 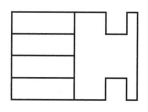

| SE Isometric | SW Isometric | Top View |

INTERSECTION
If solid objects intersect, they <u>share a space</u>. This shared space is called an Intersection. The **Intersection** command allows you to create a solid from this shared space.

1. Select the **Intersection** command using one of the following:

    Ribbon = Home tab / Solid Editing /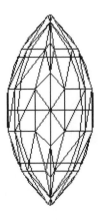

    Keyboard = in

2. Select objects: ***select the solid objects that form the intersection***
    Select objects: ***<enter>***

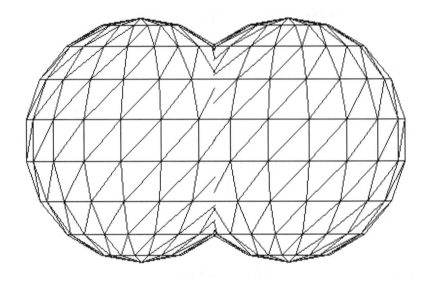

BEFORE INTERSECTION

AFTER INTERSECTION

EXERCISE 15A
Subtract

1. **Select File / New and select <u>Acad3d.dwt</u>.**

2. Select the **SE Isometric view.**

3. Draw the solid object shown below. (Step by step instructions on the next page)

4. Save as **EX-15A.**

5. **Do not Dimension.**

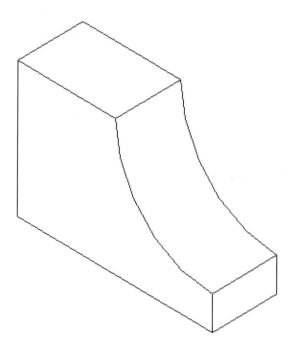

Step by step instructions and dimensions are shown on the next page.

Note: There are many different methods to create the object above.
The steps on the next page are designed to make you think about UCS
positioning, rotating and positive and negative inputs.
But you may use any method you choose.

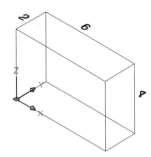

Step 1.
Create a solid box.

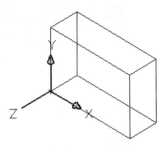

Step 2.
Rotate the UCS icon (pg. 15-6)
or 3point (pg. 15-4) or Dynamic (pg 15-5)

Cylinder
Center
Point

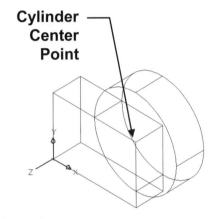

Step 3.
Draw a 3" Radius Cylinder
(Note: Ht = negative 2)

**Use 3D Orbit to confirm that you
placed the Cylinder correctly.**

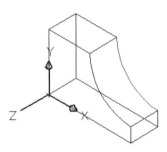

Step 4.
Subtract the Cylinder from the box

EXERCISE 15B
Union and Subtract

1. Open **EX-15A**

2. Add and subtract from the existing model to form the solid object below.
   (Step by step instructions on the next page)

3. Save as **EX-15B**

4. **Do not Dimension**

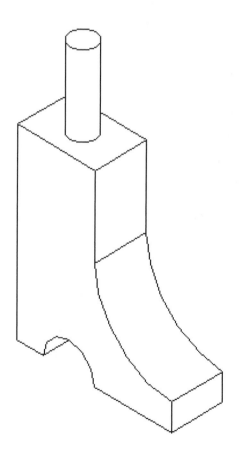

A few construction hints and dimensions are shown on the next page.

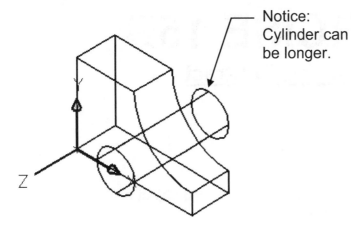

Notice:
Cylinder can
be longer.

R1.000   2.000

Step 1.
Subtract a Cylinder

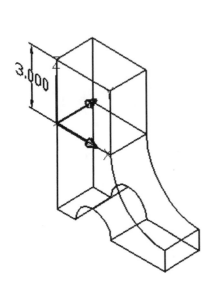

3.000

Step 2.
Add a solid Box.

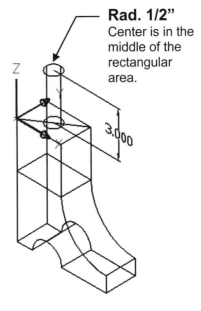

Rad. 1/2"
Center is in the
middle of the
rectangular
area.

3.000

Step 3.
Add a Cylinder.

Step 4.
Union all 3 parts

EXERCISE 15C
Assembling 3D solids

1. **Select File / New and select <u>Acad3d.dwt</u>.**

2. Change the **Units** to Architectural and the **Limits** to 60, 60 (inches).
   Reminder: Type: Units <enter>   Type: Limits <enter>

3. Draw the table shown below
        (Step by step instruction on the next page)

4. Save as **EX-15C**

5. Do not dimension

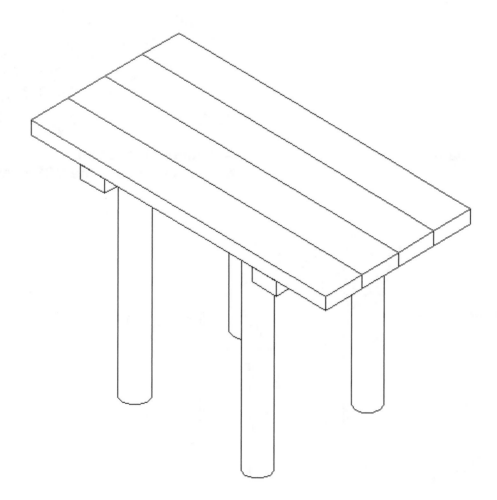

Dimensions and construction hints on the next page.
Have some fun with this.
If you have time try to add some benches.

Overall Dimensions

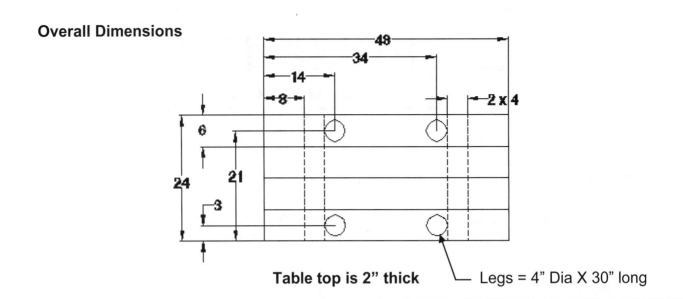

Table top is 2" thick

Legs = 4" Dia X 30" long

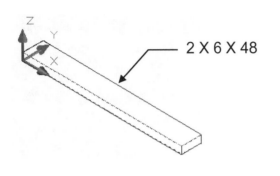

2 X 6 X 48

Step 1

Start with a plank.

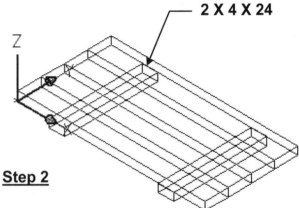

2 X 4 X 24

Step 2

Add more planks and
the 2 x 4's (Notice the UCS position)

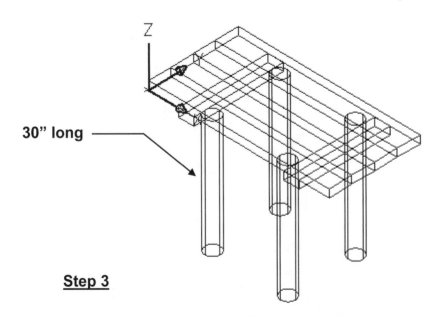

30" long

Step 3

Add the <u>30" long</u> legs
Remember, to make the legs go in the negative Z axis

LEARNING OBJECTIVES

After completing this lesson, you will be able to:

1. Extrude solid surfaces.
2. Create a Region.
3. Review Polyline "Join" command.
4. Use Presspull command.
5. Add Thickness to a surface.

Sorry LT users, you do not have this option.

LESSON 16

EXTRUDE

AutoCAD's solid objects are helpful in constructing 3D solids quickly. But what if you want to create a solid object that has a more complex shape, such as the one shown here.

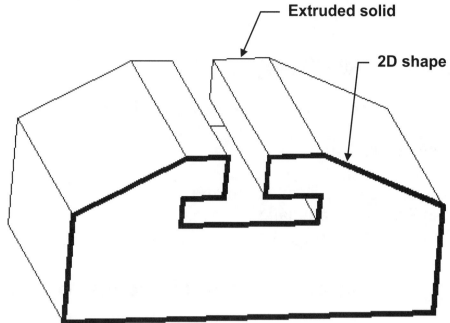

Extruded solid

2D shape

To create the solid shown above you first need to draw a **closed 2D shape** and then **EXTRUDE** it

The **EXTRUDE** command allows you to take a 2D shape and extrude it into a solid. It extrudes <u>along the Z-Axis.</u> Extrusions can be created along a straight line or along a curved path it may also have a taper. (Refer to page 16-4)

You may EXTRUDE Polylines, Arcs, Circle, Polygon, Rectangle, Ellipses, Donuts, and Regions. (Splines can also be extruded but we are not discussing those in this workbook)

4 methods for extruding a 2D shape:
1. Perpendicular to the 2D shape with straight sides. (page 16-3)
2. Perpendicular to the 2D shape with tapered side. (page 16-4)
3. Along a Path (Page 16-4)
4. Extrude a Region. (Page 16-5)

HOW TO SELECT THE EXTRUDE COMMAND

Ribbon = Home tab / Modeling panel /

Keyboard = Ext

EXTRUDE - <u>PERPENDICULAR WITH STRAIGHT SIDES</u>

Before drawing the 2D shape do the following:

1. Select the **Top** view.

2. Draw the closed 2D shape. (Be sure to use "close")

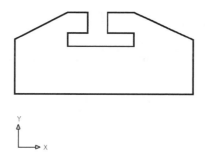

Now Extrude:

1. Select the **SE Isometric view.**

2. Select the **EXTRUDE** command using one of the methods shown on page 16-2.

3. Select the objects: *select the objects to extrude <enter>*

4. Select the objects: *<enter>*

5. Specify height of extrusion or [Direction/Path/Taper angle]: *type the height <enter>*

   (Note: a **positive** value extrudes **above the XY plane** and **negative** extrudes **below the XY plane,** along the Z axis.)

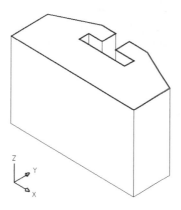

EXTRUDE - <u>PERPENDICULAR WITH TAPERED SIDES</u>

1. Draw a closed 2D shape on the XY plane.
2. Select the **EXTRUDE** command
3. Select the objects: ***select the objects to extrude \<enter>***
4. Select the objects: ***\<enter>***
5. Specify height of extrusion or [Direction/Path/Taper angle]: ***T \<enter>***
6. Specify angle of taper for extrusion <0>: ***enter taper angle \<enter>***
7. Specify height of extrusion or [Direction/Path/Taper angle]: ***enter height \<enter>***

How to control the taper direction
If you enter a <u>positive angle</u> the resulting extruded solid will <u>taper inwards</u>.
If you enter a <u>negative angle</u> the resulting extruded solid will <u>taper outwards</u>.

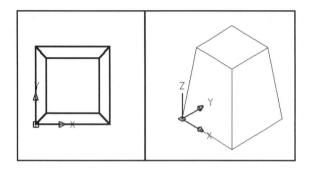

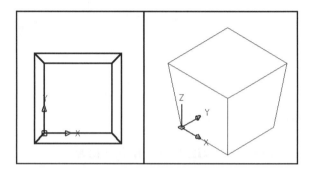

Positive angle   **Negative angle**

EXTRUDE - <u>ALONG A PATH</u>

Important: The path should be drawn perpendicular to the 2D shape to be extruded.

1. Draw a 2D shape on the XY plane.
2. Draw the path perpendicular to the shape.
   (hint: rotate the UCS 90 degrees)
3. Select the **EXTRUDE** command.
4. Select the objects: ***select the objects to extrude \<enter>***
5. Specify height of extrusion or [Direction/Path/Taper angle]: ***P \<enter>***
6. Select extrusion path or [Taper angle]:***select the path (object)***

Notice the path must be drawn perpendicular to the shape to be extruded. Draw the shape then move the XY plane perpendicular to the shape.

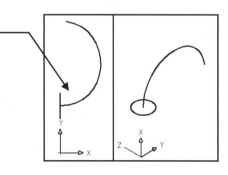

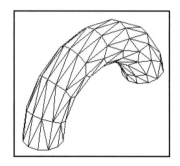

REGION

A **REGION** is a solid with no thickness. Think of a piece of paper. Thin but it is a solid. You can use all the Boolean operations, Union, Subtract and Intersect, on the region and you can extrude it.

Example:

1. You have a **2D drawing of a flat plate with circles** on it. You would like to extrude this plate and you want the circles to be actual holes.

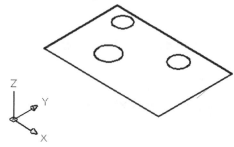

2. Use the **REGION** command to transform these objects into a solid with holes.

   a. Select the **REGION** command using one of the following:
      Ribbon = Home tab / Draw panel ▼ /

      Keyboard = reg

   b. Select objects: ***select the objects \<enter>***
      Select objects: ***\<enter>***
      4 loops extracted.
      4 Regions created.

3. Now **subtract** the circles from the rectangle, so they are actually "holes".

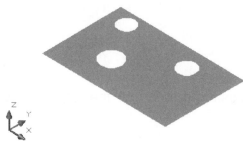

4. Extrude the region.

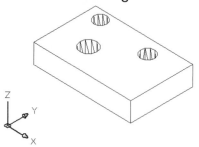

Which is more efficient?

1. Draw a 2D drawing, create a region and then extrude.
      or
2. Draw a box and 3 cylinders then subtract cylinders from box.

Also see the PressPull command on page 16-6.

PRESSPULL COMMAND

The **PressPull** command allows you to create a Region and Extrude a closed shape all in one operation. You may also use it to modify the height of an already extruded object. The area must be closed such as a circle, rectangle, polygon or closed polylines. You may even select an area created by two overlapping objects.

1. Draw a shape. (Note: it is not necessary to "subtract" the circles)

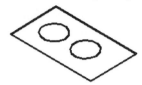

2. Move the UCS to the plane to extrude if the objects are not on the XY plane. (The plane to be extruded must be on the XY plane.

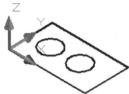

3. Select the Press/Pull command using:

 Ribbon = Home tab / Modeling /

 Keyboard = PressPull

4. Click in the boundary just as you would to select a hatch boundary.

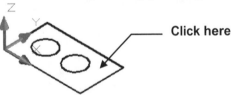

Click here

5. Move the cursor along the positive or negative Z axis to press or pull the region. You may click to determine the height or enter a value.

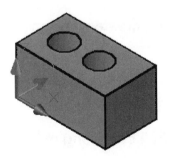

POLYSOLID

If you Extrude an open shape you will create 3D surfaces instead of a solid. These surfaces have no thickness. Sort of like a piece of paper standing on end. You may give these surfaces thickness using the Polysolid Surface tool.

1. Select the Polysolid using one of the following:

    Ribbon = Home tab / Modeling /

    Keyboard = polysolid

    Command: _Polysolid Height = 8'-0", Width = 0'-4", Justification = Center
2. Specify start point or [Object/Height/Width/Justify] <Object>: *Type h <enter>*

3. Specify height <8'-0">: *Type the height desired <enter>*

    Height = 8'-0", Width = 0'-4", Justification = Center

4. Specify start point or [Object/Height/Width/Justify] <Object>: *Type w <enter>*

5. Specify width <0'-4">: *Type the width desired <enter>*

    Height = 8'-0", Width = 0'-4", Justification = Center

6. Specify start point or [Object/Height/Width/Justify] <Object>: *Type j <enter>*

7. Enter justification [Left/Center/Right] <Center>: *Type r <enter>*

    Height = 8'-0", Width = 0'-4", Justification = Right

8. Specify start point or [Object/Height/Width/Justify] <Object>: *Place the start point*

9. Specify next point or [Arc/Undo]: *Place next point*

10. Specify next point or [Arc/Undo]: *Place next point*

11. Specify next point or [Arc/Undo]: *continue placing next points or <enter> to stop*

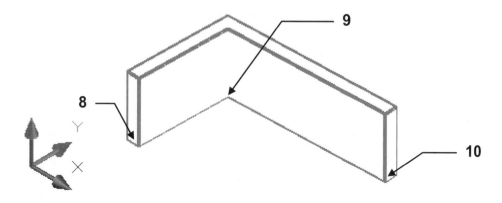

DELOBJ SYSTEM VARIABLE

The **DELOBJ** system variable determines whether the object used when extruding, revolving, sweeping or presspulling, is retained or deleted. For example, when you extrude a circle to make a cylinder, what should happen to the original circle? Do you want to delete it or do you want to keep it?

Set the **DELOBJ** system variable before using the Modeling command.

1. To set the variable type **delobj <enter>**

2. Enter one of the options described below.

Options:

0 = Retains all source objects.

1 = This is the default setting. Deletes profile curves. For example, if you use a circle to create a cylinder, the circle will be deleted. Cross sections and guides used with the SWEEP command are also deleted. However, if you extrude along a path, the path is not deleted.

2 = Deletes all defining objects, including paths

-1 = Just like "1" except you get a prompt to allow you to choose.

-2 = Just like "2" except you get a prompt to allow you to choose.

PLAN VIEW

Remember, you always draw on the XY plane. If you want to draw on a specific surface you have to move the XY plane (Origin) to the surface that you wish to work on. Then you may position the view in a 2D plan view so you may add a feature easily. You may do this easily with the PLAN command. After you move the XY Plane (Origin) to the surface type "Plan <enter>" and the view will change to a 2D plan view of the surface.

1. Move the UCS icon to the plane that you wish to draw on. (Refer to page 15-4)
2. Type: **PLAN <enter>**
3. Enter an option [Current ucs/Ucs/World] <Current>: **<enter>**

The "PLAN" viewport will display the current XY plane.
You can now draw on the XY plane as if you were drawing a 2D drawing.

> **Very Important:**
> **"DUCS must be OFF" to use this method of drawing.**

EXAMPLE

Step 1. Move Origin

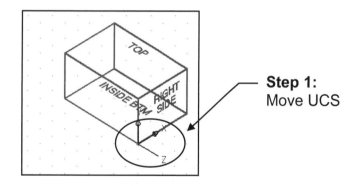

Step 1:
Move UCS

Step 2. Type Plan <enter>

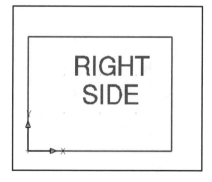

Now you may add features easily as if you were drawing in 2D

EXERCISE 16A
Extrude

1. Start a **New** file and select **Acad3d.dwt.**
2. Draw the 2D drawing approximately as shown below to form a closed Polyline. (Fig. 1)
   (Exact size is not important.)
3. Use Extrude or Presspull to form the solid object. Height = 4" (Fig. 2)
4. Select the **SE Isometric** view.
5. Save as **EX-16A**

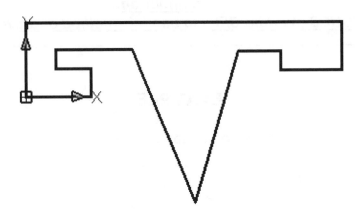

Figure 1

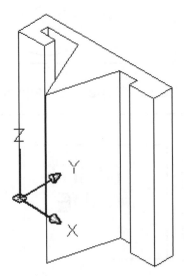

Figure 2

EXERCISE 16B
Extrude with Taper

1. Start a **New** file and select **Acad3d.dwt.**
2. Draw a 6 sided Inscribed Polygon with a radius of 2 (Fig. 1)
3. Extrude to form the solid object. Ht = 4 Taper = 5 (Fig. 2)
4. Add the letter & Number shapes (Use polyline, **do not use text**).
   Extrude each. Ht = 1 (Fig. 3)
   Don't forget to use subtract on the shapes "4" , "6" and "B".

Refer to the next page for construction suggestions

5. Select the **SE Isometric** view.
6. Save as **EX-16B**

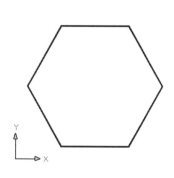

Figure 1

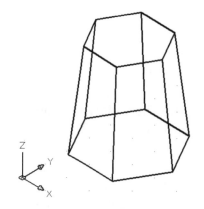

Figure 2

Figure 3

Construction suggestion:

a. After constructing the shape shown below, move the UCS, using 3 Point (pg 15-3), to select the correct surface to draw on.

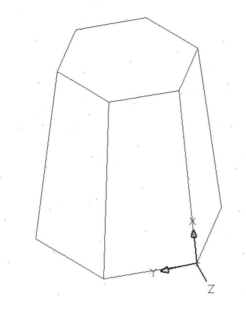

b. View the **Plan view** to draw the text.
   (Refer to page 16-9)

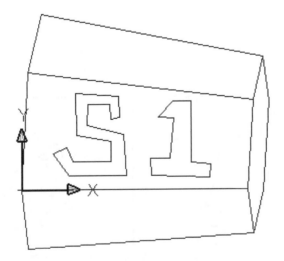

c. Extrude the polyline letter shapes.

d. Select **SE Isometric view** and then rotate around to the next surface.

EXERCISE 16C
Extrude along a Path

1. Start a **New** file and select **Acad3d.dwt.**
2. Draw a 6 sided Inscribed Polygon with a radius of 1 (Fig. 1)
3. Rotate the UCS, about the Y axis, -90. (Perpendicular to current UCS.)
4. Draw a **polyline path** perpendicular to the Polygon approximately as shown (Fig. 2)
5. Extrude the Polygon along the path (Fig. 3)
6. Select the **SE Isometric** view.
7. Save as **EX-16C**

Path

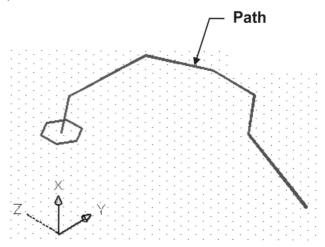

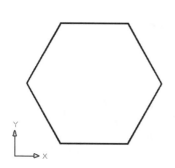

Figure 1

Figure 2

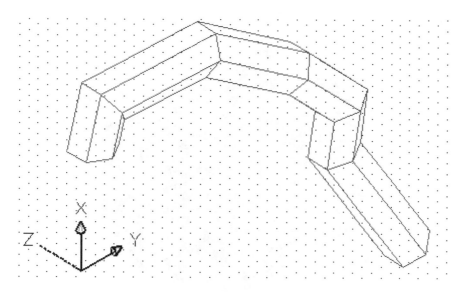

Figure 3

EXERCISE 16D
Extrude or PressPull a Region

1. Start a **New** file and select **Acad3d.dwt.**
2. Draw the 2D drawing shown below. (Figure. 1)
   Note: Use **PLAN**
3. Create a "Region" (pg. 16-5) and Extrude or use "PressPull" command
   to form the solid object (Figure. 2).   Height = 4"
4. Save as **EX-16D**

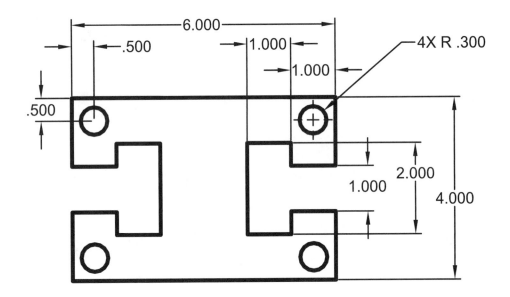

Figure 1

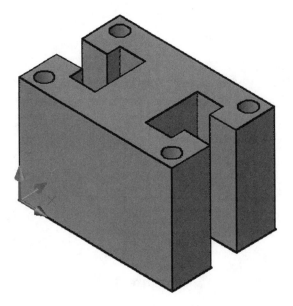

Figure 2

EXERCISE 16E
Extrude or PressPull a Region

1. Open **EX-5D**
2. **Freeze** all layers except "Walls". (Hopefully you used the correct layers when drawing 5D.)
3. Select the **SE Isometric** View.
4. Use the "**REGION**" command (refer to page 16-5) to create 4 regions.

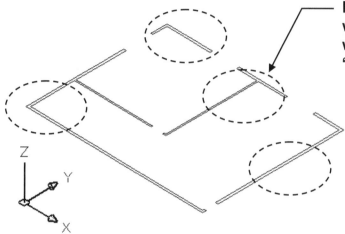

Note: The missing doors and windows cause a separation in the walls. You must use "Join" or "Region" on each wall section.

5. EXTRUDE or Presspull the walls to a height of 8 feet.
6. Save as **EX-16E**

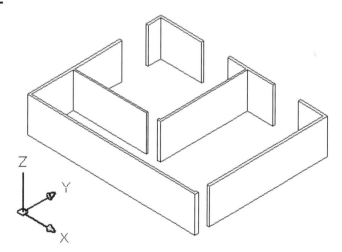

To add walls or windows you can use the Union or Subtract. Have some fun with this one. Try adding the furniture.

NOTES:

<u>**LEARNING OBJECTIVES**</u>

After completing this lesson, you will be able to:

1. Understand 3D Operations.
   Mirror 3D, 3D Rotate, 3D Align and 3D Array

LESSON 17

3D OPERATIONS

3D Operations allow you to **3D Mirror, 3D Rotate, 3D Align** and **3D Array** a solid. The methods are almost identical to the 2D commands with the exception of defining the plane. <u>You do not have to move the UCS.</u>

Note: You may use the equivalent 2D commands but they will only work in the XY plane. <u>It may be necessary to move the UCS.</u>

THE FOLLOWING ARE EXPLANATIONS FOR THE <u>3D COMMANDS.</u>

3D MIRROR

You must define the mirror plane.

1. Select the **3D MIRROR** command using one of the following:

    Ribbon = Home tab / Modify panel /

    Keyboard = 3dmirror or mirror3d

2. Select objects: *select the solid to be mirrored*

3. Select objects: *Select more or <enter> to stop*

4. Specify first point of mirror plane (3 points) or [Object/Last/Zaxis/View/XY/YZ/ZX/3points] <3points>: *3<enter>*

5. Specify first point on mirror plane: *select first point on mirror plane (P1)*

6. Specify second point on mirror plane: *select second point on mirror plane (P2)*

7. Specify third point on mirror plane: *select third point on mirror plane (P3)*

8. Delete source objects? [Yes/No] <N>:*select Y or N <enter> (the example is NO)*

EXAMPLE:

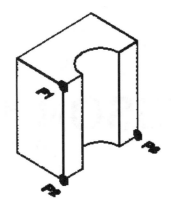
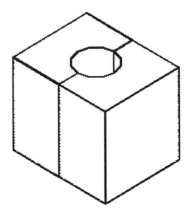

3D ROTATE

You must pick 2 points to define the axis of rotation and the rotation angle. To determine the rotation angle you must look down the axis from the second point.

If you use my stabbing arrow analogy (page 15-6), you will put the positive end of the second point selected in your chest and rotate the other two axes. *Positive input is counterclockwise and Negative input is clockwise.*

1. Select the **3D ROTATE** command using one of the following:

   Ribbon = Home tab / Modify panel /

   Keyboard = 3drotate

2. Current positive angle in UCS: ANGDIR=counterclockwise ANGBASE=0
   Select objects: ***select the solid to be rotated***

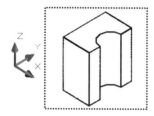

3. Select objects: ***<enter> to stop***

4. Specify Base Point: ***Snap to the base point of the rotation***

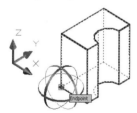

5. Pick a Rotation axis:
   Touch the ribbons until the correct axis line appears then click left mouse button

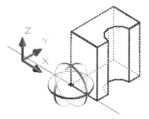

In the example on the left I have selected the Red ribbon which is the "X" axis. A Red line should appear.

6. Specify angle start point: ***type rotation angle <enter>***

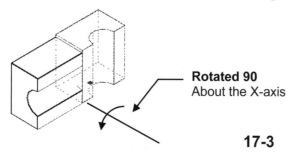

Rotated 90
About the X-axis

17-3

3D ALIGN

The **3D ALIGN** command allows you to MOVE and ROTATE an object from its existing location to a new location.

You define the two or three points on the **SOURCE** object and then the points on the **DESTINATION** location.

1. Select the **3D ALIGN** command using one of the following:

    Ribbon = Home tab / Modify panel /

    Keyboard = 3dalign

2. Select objects: *select the object to be aligned*

3. Select objects: *select more or <enter> to stop*

4. Specify source plane and orientation ...
    Specify base point or [Copy]:: *select the 1st source (S1)*

5. Specify second point or [Continue] <C>: *select the 2nd source (S2)*

6. Specify third point or [Continue] <C>: *select the 3rd source (S3)*

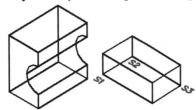
It will be easier to select source points if you change the "**Visual Style**" to **3D Wireframe**. (Refer to page 13-11)

7. Specify destination plane and orientation ...
    Specify first destination point: *select the 1st destination point (D1)*

8. Specify second destination point or [eXit] <X>: *select the 2nd destination point (D2)*

9. Specify third destination point or [eXit] <X>: *select the 3rd destination point (D3)*

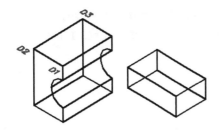

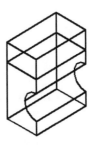

Completed alignment

3D ARRAY (Rectangular and Polar)

RECTANGULAR ARRAY

You can array any object on the current XY Plane.
You define the following:

ROWS = number of copies needed in the **Y** direction
COLUMNS = number of copies needed in the **X** direction
LEVELS = number of copies needed in the **Z** direction
DISTANCE BETWEEN ROWS, COLUMNS and LEVELS

1. Select the **3DARRAY** command using one of the following:

   Ribbon = Home tab / Modify panel /

   Keyboard = 3darray

2. Select objects: *select object to be arrayed*

3. Enter the type of array [Rectangular/Polar] <R>: *R <enter>*

5. Enter the number of rows (---) <1>: *type number of rows (Ydirection) <enter>*

6. Enter the number of columns (||||) <1>: *type number of columns (X direction) <enter>*

7. Enter the number of levels (...) <1>: *type number of levels (Z direction) <enter>*

8. Specify the distance between rows (---): *type distance between rows<enter>*

9. Specify the distance between columns (||||): *type distance between columns<enter>*

10. Specify the distance between levels (...): *type distance between levels <enter>*

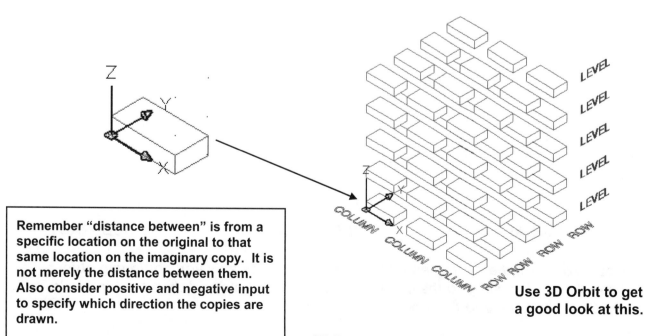

Remember "distance between" is from a specific location on the original to that same location on the imaginary copy. It is not merely the distance between them. Also consider positive and negative input to specify which direction the copies are drawn.

Use 3D Orbit to get a good look at this.

17-5

POLAR ARRAY

The object is arrayed around an <u>entire axis</u> rather than just a point.

You define the following:
> **NUMBER OF COPIES**
> **ANGLE TO FILL**
> **ROTATION DIRECTION**
> **ROTATION AXIS ENDPOINTS**

1. Select the **3D ARRAY** command using one of the following:

    Ribbon = Home tab / Modify panel /

    Keyboard = 3darray

2. Select objects: *select object to be arrayed*

3. Select objects: *<enter>*

4. Enter the type of array [Rectangular/Polar] <R>: *P <enter>*

5. Enter the number of items in the array: *type the number of copies (include original)*

6. Specify the angle to fill (+=ccw, -=cw) <360>: *enter angle for copies to fill*

7. Rotate arrayed objects? [Yes/No] <Y>: *copies rotated, yes or no?*

8. Specify center point of array: <Osnap on>*snap to first endpoint of axis (P1)*

9. Specify second point on axis of rotation:*snap to second endpoint of axis (P2)*

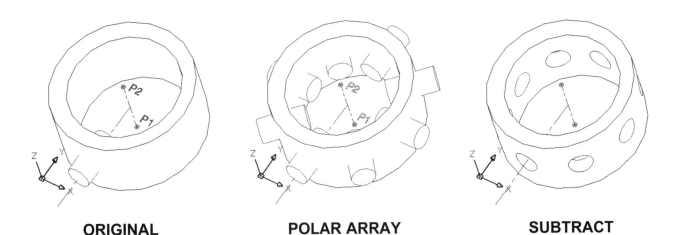

| ORIGINAL | POLAR ARRAY | SUBTRACT |

Note: To select the 1st (P1) and 2nd (P2) points on the axis you could use object snap "center" to snap to the center of the bottom circle and then the top circle. Just want to keep you thinking.

EXERCISE 17A
3D MIRROR

1. Start a **New** file and select **Acad3d.dwt.**
2. Draw the solid object below.
3. Dimensions and construction hints on page 17-8.
4. Save as **EX-17A**

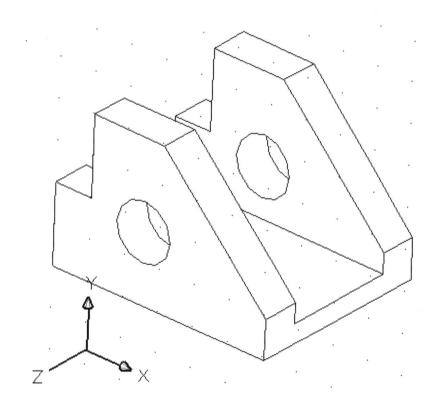

Drawing hints on the next page.

Step 1. Draw the side profile using Plan view.

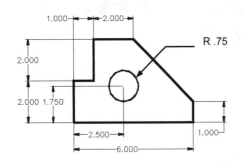

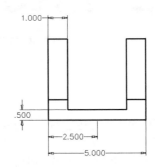

Don't forget to use Polyline and Join if necessary.

R .75

1.000 2.000 2.000 1.750 2.500 6.000

1.000 .500 2.500 5.000

Step 2. Presspull (1.00)

Step 3. Add center divider. (L=6 W=.50 H=1.50)

Step 4. Rotate

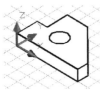

Step 5. Mirror

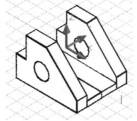

Step 6. Union

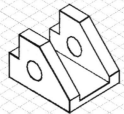

EXERCISE 17B
3D ROTATE

1. Open **EX-17A** (Figure 1)
2. Rotate the solid model 180 degrees as shown in Figure 2
3. Save as **EX-17B**

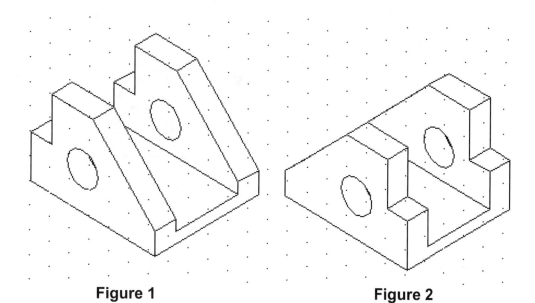

Figure 1   Figure 2

EXERCISE 17C
3D ALIGN

1. Start a **New** file and select **<u>Acad3d.dwt</u>**.
2. Draw the 3 solid objects shown below in Figure 1
   (Use the dimensions shown)
3. Using the **3D ALIGN** command, assemble the objects as shown in Figure 2.
4. Save as **EX-17C**

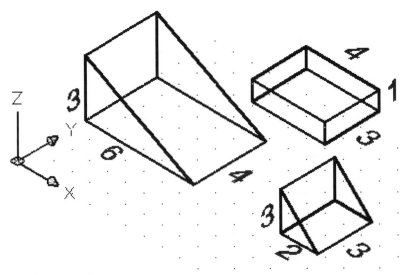

Figure 1

Note: It will be easier to select the source points if you change the Visual Style to 3D Wireframe

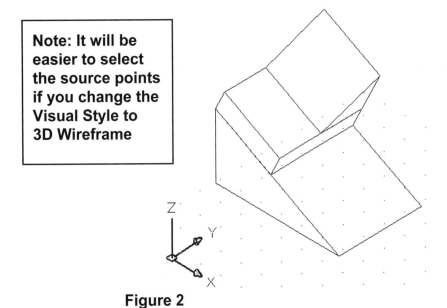

Figure 2

EXERCISE 17D
2D ARRAY

1. Start a **New** file and select **<u>Acad3d.dwt</u>**.
2. Draw the box shown below in Figure 1 (Use the dimensions in Figure 2).
3. Try the **2D** Array command to draw the holes.
4. Create a Region, Subtract the circles and Extrude. (Refer to lesson 16) Or select any method you prefer.
5. Save as **EX-17D.**

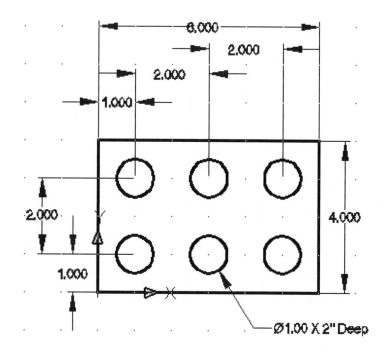

Figure 1

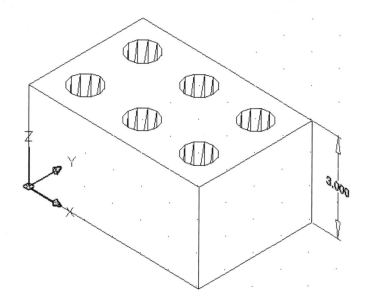

Figure 2

EXERCISE 17E
3D ARRAY - RECTANGULAR

1. Start a **New** file and select **Acad3d.dwt.**
2. Draw the solid objects shown below in Figure 1
   (Use the dimensions shown at the bottom of the page)
3. Try the **3D Array** command to create the solid model in Figure 2.
4. Save as **EX-17E**

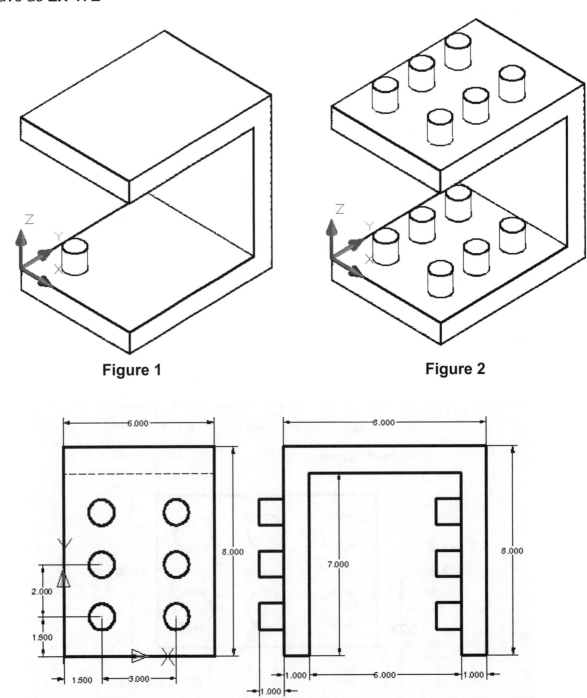

Figure 1   **Figure 2**

EXERCISE 17F
3D ARRAY - POLAR

1. **Select File / New and select <u>Acad3d.dwt</u>.**
2. Create the solid cylinder with 8 holes shown below
    (Dimensions and construction hints on page the next page)
3. Save as **EX-17F**

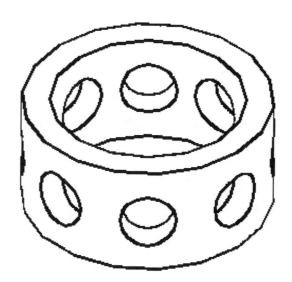

Drawing hints on the next page and refer to page 17-6

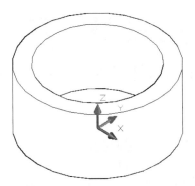

Step 1. Draw 2 Cylinders. 9" Dia and 7" Dia. Height = 4"
Subtract the small from the large.

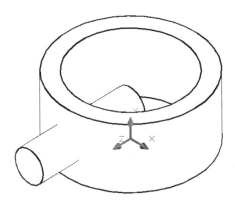

a. Rotate the UCS
b. Draw Cylinder 2"
up from the bottom.
 Height = 7"
(You really have to
think about this)

Step 2. Draw a 2" Dia Cylinder

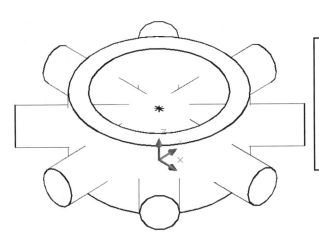

a. Rotate the UCS
b. Polar Array
around the UCS.
Hint:
First pt = 0,0,0
Second pt = 0,0,4

Step 3. Polar Array 8 Cylinders

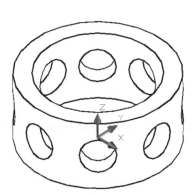

Step 4. Subtract the small cylinders from the large Cylinder.

LEARNING OBJECTIVES

After completing this lesson, you will be able to:

1. Use 3D grips to modify Solid objects

LESSON 18

STRETCH a 3D Solid using 3D GRIPS

3D Grips will make modifying your 3D model very easy. Click on any Single 3D object and the grips will appear.

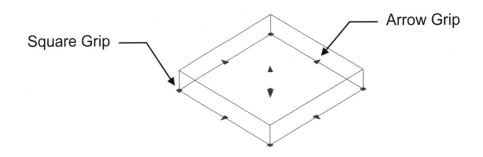

Square Grip

Arrow Grip

If you click on a square you can adjust both intersecting sides. (Ortho off)

If you click on an arrow you can adjust an individual side, top or bottom.

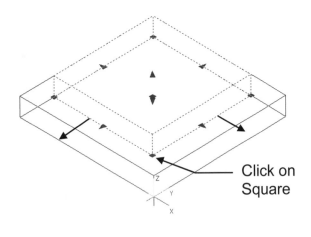

Click on Square

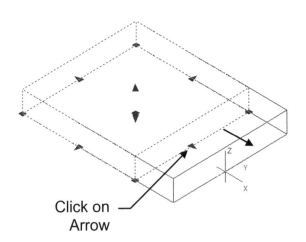

Click on Arrow

If **DYN** is **ON** while you adjust the sides you may enter a distance relative to the original location.

STRETCH a 3D Solid with subobjects using 3D GRIPS.

If you click on a 3D solid that contains subobjects, such as a hole, only the move grip will appear.

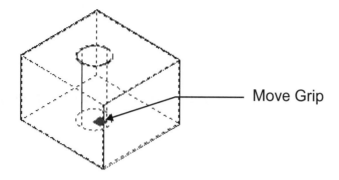

— Move Grip

To modify the individual subobjects you must use the Ctrl key.

1. Hold the **Ctrl** key down and **click** on the individual subobject.

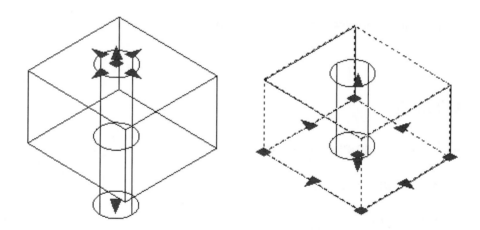

2. Adjust the size using the grips as shown on the previous page.

MOVE 3D SOLIDS using GRIPS or 3D MOVE

You may easily move a 3D solid using the move grip or the 3D Move command.

MOVE GRIP

1. Click on the 3D solid and the grips will appear.

Notice the 3D object with a subobject on the right has only one grip.

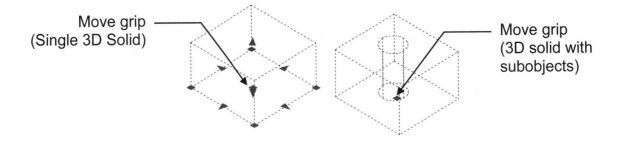

Move grip
(Single 3D Solid)

Move grip
(3D solid with
subobjects)

2. Click on the move grip and drag to new location.

3D MOVE COMMAND

1. Select the 3D MOVE command using one of the following:

   Ribbon = Home tab / Modify panel /

   Keyboard = 3dmove

2. Select the object: *select the object to be moved*

3. Select the object: *select more objects or <enter> to stop*

4. Specify base point or [Displacement] <Displacement> *select the basepoint*

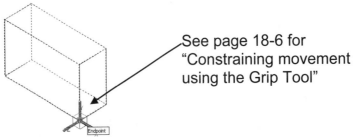

See page 18-6 for
"Constraining movement
using the Grip Tool"

5. Specify second point or <use first point as displacement>: *place new location*

MOVE A <u>SUBOBJECT</u> using GRIPS

The following will show you how to move a subobject such as a hole.

Note: The appearance of the Grips depends on how the subobject was created.

1. Select the hole as follows:
   - place the cursor <u>inside the hole</u> and <u>Crtl+Click</u>.

Grip appearance if the hole was created using: "**Subtract**"

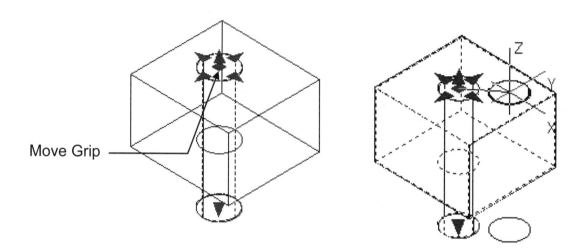

Move Grip

Grip appearance if the hole was created using:
Extrude or **PressPull** a surface

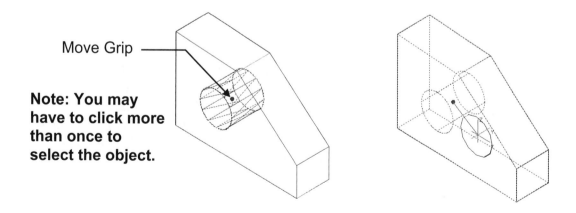

Move Grip

Note: You may have to click more than once to select the object.

2. Click on the move grip and drag the hole to the new location.

3. Click to fix the new location.

Constrain the Movement using the Gizmo.

You may constrain the movement to one axis or and entire plane such as XY.
Note: you may also use Ortho if you are constraining the movement to just one axis.

1. Click on the 3D solid and the grips will appear as shown on page 18-5.

The "**Gizmo**" will appear.

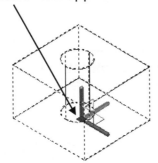

*Note: **Gizmo** will not appear if visual style is 2D wireframe.*

2. Rest the cursor on the X-axis arm (red) of the tool (do not click yet). A red line appears and extends across the drawing area and the Gizmo X-axis arm changes color.

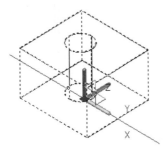

3. Now click on the X-axis arm and move the cursor. You will notice that the movement is constrained in the X-axis.

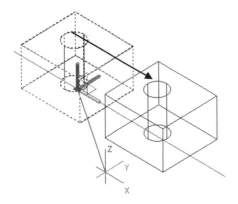

Note: You may restrict the movement to an entire plane, such as XY, if you place the cursor between 2 axis arms until they both change color then click.

4. Click to establish the new location. (Press the Esc key to clear the grips.)

EXERCISE 18A
Create a Cube

1. Start a **New** file and select **Acad3d.dwt**

2. Draw a 3D Solid Cube as shown below.

3. Save as: **EX-18A**

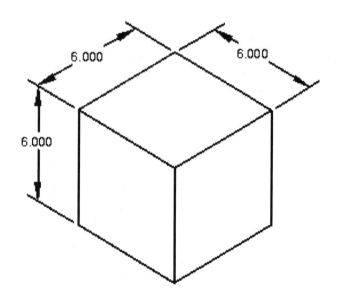

EXERCISE 18B
PressPull or Grips

1. Open Ex-18A

2. Increase the Length and Height as shown below.

3. Save as: EX-18B

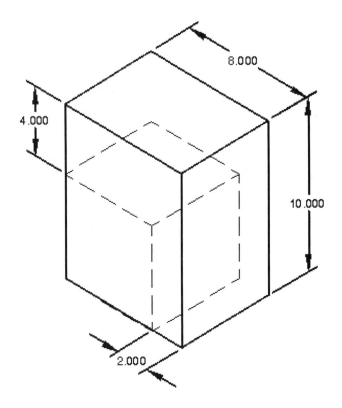

EXERCISE 18C
Add Cylinders and Subtract

1. Open Ex-18B

2. Add (3) 1" radius holes as shown below.

3. Save as: EX-18C

Notice:
You will have to move the Origin and change the XY plane
to accurately place the holes.

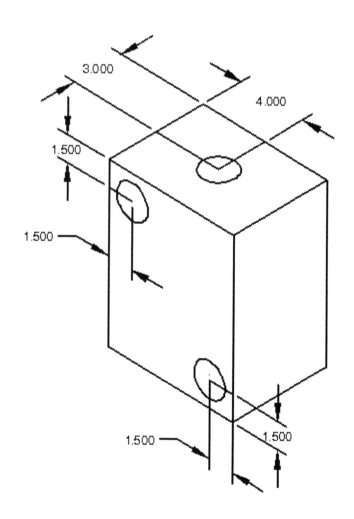

EXERCISE 18D
Grips or 3D Move

1. Open Ex-18C

2. Move hole "A" to a new location as shown below.

3. Save as: EX-18D

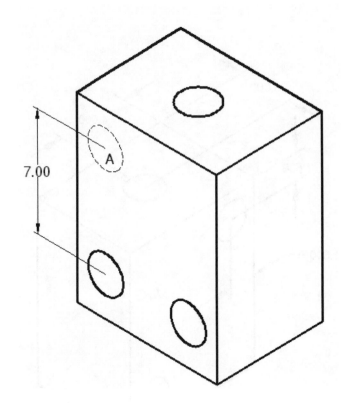

7.00

A

EXERCISE 18E
Properties, Grips or 3D Offset

1. Open Ex-18D

2. Increase the size of the hole "B" to 2" radius.

3. Save as: EX-18E

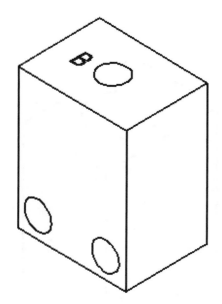 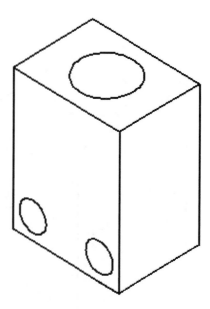

EXERCISE 18F
Grips or 3D Delete

1. Open Ex-18E

2. Delete hole "C".

3. Save as: EX-18F

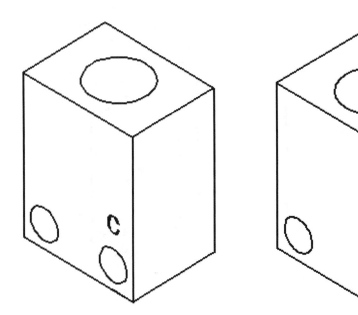

LEARNING OBJECTIVES

After completing this lesson, you will be able to:

1. Create a solid object by revolving a 2D shape
2. Slice a solid object into 2 segments
3. Create a 2D and 3D section view through a solid object.
4. Create a twisted object along a path.
5. Create a Helix.

LESSON 19

REVOLVE

The **REVOLVE** command allows you to revolve a profile around an axis. If the shape is closed it will become a solid. If the shape is open it will become a surface. The shape can be revolved from 1 to 360 degrees.

You will select the axis to revolve around, the solid to be revolved and enter the angle of revolution.

Selecting the axis of revolution
The axis of revolution can be the X or Y axis. It may also be an object such as a line. The axis may be located on the shape or not.

HOW TO USE THE REVOLVE COMMAND

1. Select the **REVOLVE** command using one of the following:

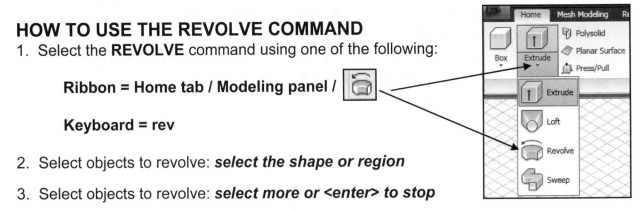

   Ribbon = Home tab / Modeling panel /

   Keyboard = rev

2. Select objects to revolve: *select the shape or region*

3. Select objects to revolve: *select more or <enter> to stop*

4. Specify axis start point or define axis by [Object/X/Y/Z] <Object>: *Select Axis*

5. Specify angle of revolution or [STart angle] <360>: *type the angle (positive ccw or negative cw determines direction of rotation.*

Below are examples of how you can make 3 different solids by selecting different "Axis of Revolution".

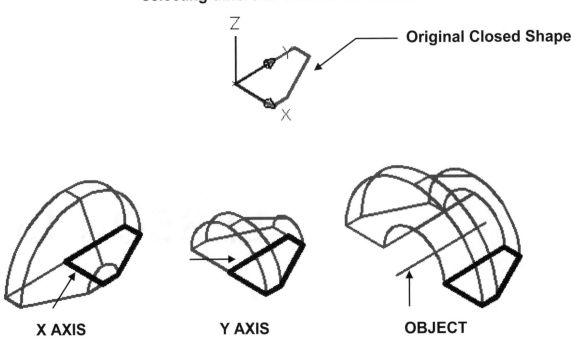

Original Closed Shape

X AXIS   **Y AXIS**   **OBJECT**

SLICE

The **SLICE** command allows you to make a knife cut through a solid. You specify the location of the slice by specifying the plane. After the solid has been sliced, you determine which portion of the slice you want removed. You may also keep both portions but they are still separate.

HOW TO USE THE SLICE COMMAND
1. Select the **SLICE** command using one of the following:

    Ribbon = Home tab / Solid editing panel /

    Keyboard = sl

2. Select objects: *select the solid object*
3. Select objects: *<enter> to stop*
4. Specify start point of slicing plane or [planer object/Surface/Zaxis/View/XY/YZ/ZX/3pts <3points>: *select the option preferred. (See methods below.)*

THERE ARE 2 EASY METHODS TO SPECIFY THE CUTTING PLANE.

Method 1. (3 points)
1. Define the cutting plane by selecting 3 points on the object.
2. Click on the side you want to keep or "B" to keep both sides.

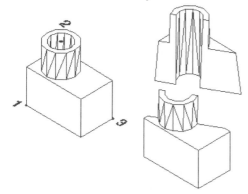

Method 2. (XY, YZ or ZX Plane)
1. Move the UCS to the desired cutting plane location.
2. Select the XY, YZ or ZX plane. (The cut will be made along the selected plane)

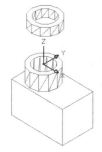 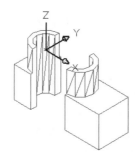 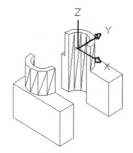

XY PLANE   **YZ PLANE**   **ZX PLANE**

SECTION

The **SECTION** command process is very similar to the Slice command. But the Section command does not actually separate the solid. It creates a 2D region from the plane that you specify. The 2D region will be created on the current layer. The 2D Region can then be moved, extruded, revolved, copied, etc.

To specify the section plane refer to the Slice methods on page 19-3.

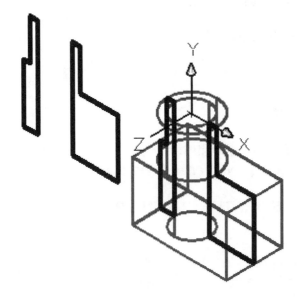

HOW TO USE THE SECTION COMMAND
1. Select the <u>Layer</u> you want the section to be on and the <u>Linetype</u>.
2. Select the **SECTION** command using one of the following:

> **Ribbon = None**
>
> **Keyboard = sec**

3. Select objects: *select the solid object*
4. Select objects: *<enter>*
5. Specify first point on section plane by [Object/Zaxis/View/XY/YZ/ZX/3points] <3points>: *select the option preferred*
   (This creates the section "in place. Now you have to move. See below)

HOW TO CREATE A HATCHED SECTION VIEW
1. Create a section (2D region) as described above.
2. Move or copy the region to a new location.
3. Explode if necessary to create separate regions.
4. Use the Hatch command to create section lines.
   (Note: The area to be hatched must be parallel to the XY plane or you will receive an error message. *My suggestion is, if you intend to hatch a section, use the XY plane method for defining the plane.)*
5. Draw any connecting lines to complete the view.

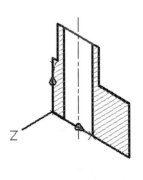

SECTION PLANE

The **SECTION PLANE** command creates a section "**object**" that acts as a "Cutting plane" through the 3D model. AutoCAD calls this command "**Live sectioning**". It allows you to cut a section plane through a model at any location, move the section plane to a new location and delete the section plane. The model will not be affected and will mend as you move the "section plane object" and the model will become whole again when you delete the section plane object.

This is a great tool to dynamically peek inside your model as you are creating it.

Examples:

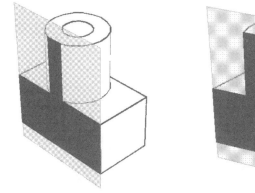 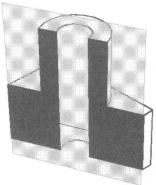 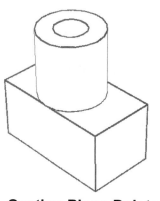

| Section Plane | Section Plane Rotated | Section Plane Deleted |

1. Select the **SECTIONPLANE** command using one of the following:

   Ribbon = Home tab / Section panel /
   Keyboard = sectionplane

2. _Sectionplane   Select Face or any point to locate section line or [Draw section / Orthographic]: *place the cursor on any "face" and click.*

3. Now you may click on the "Section Plane object" and move it.

After selecting the Section Plane you may move or rotate it using the "Arrow grip" or "Square grip".

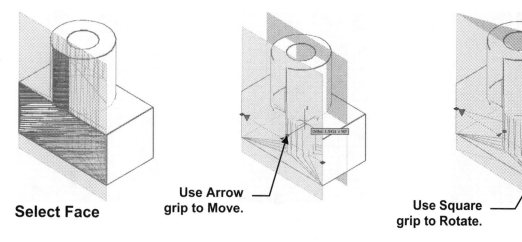

Select Face   **Use Arrow grip to Move.**   **Use Square grip to Rotate.**

SECTION PLANE....continued

When you have successfully placed the Section Plane object you may do the following:

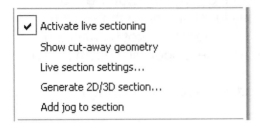

Note: Left click on the Section Plane object and then Right click. A Pop up menu will appear that includes the options shown on the left.

Activate live sectioning: Activate or De-Activate the live sectioning.

Show cut-away geometry: displays cut-away geometry

Live section settings: refer to the next page

Generate 2D/3D section... This is very useful. This option will create either a 2D or 3D section from the Area that the Section Plane object defines.

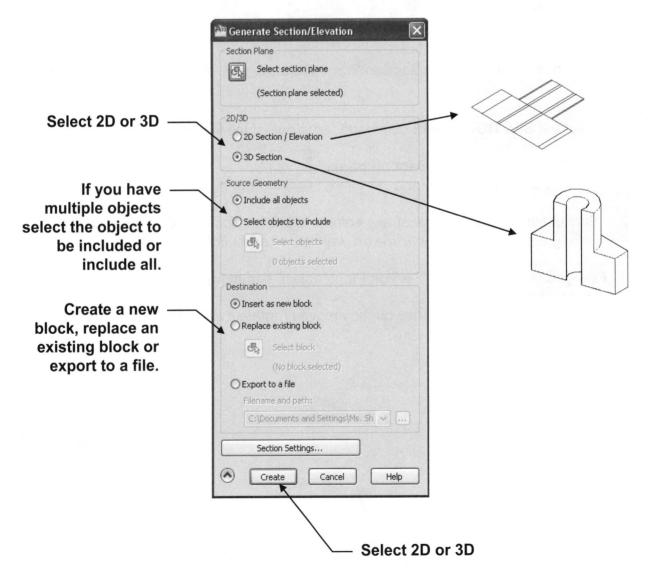

Add Jog to section: Specify a location for the jog by clicking on the section line.

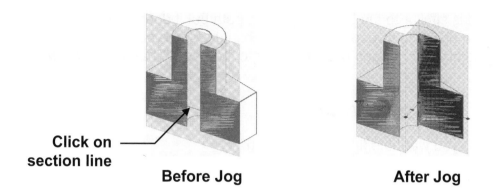

Click on section line

Before Jog   **After Jog**

Live section settings.... Determines how sectioned objects are displayed. You may make changes to the "Properties" of the Section Plane, 2D section or 3D section.

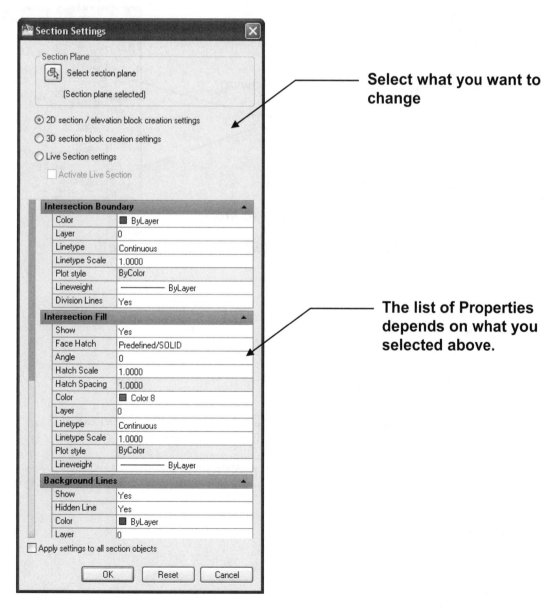

Select what you want to change

The list of Properties depends on what you selected above.

SWEEP

The SWEEP command is similar to the Extrude along a path but it does much more. You draw a path and an object to sweep along that path in the same plane. It will automatically aligns the object to sweep perpendicular to the path. That makes drawing them a lot easier.

HOW TO USE THE SWEEP COMMAND

1. Select the Layer that you want the finished object on.

2. Draw the object and the path on the same plane.

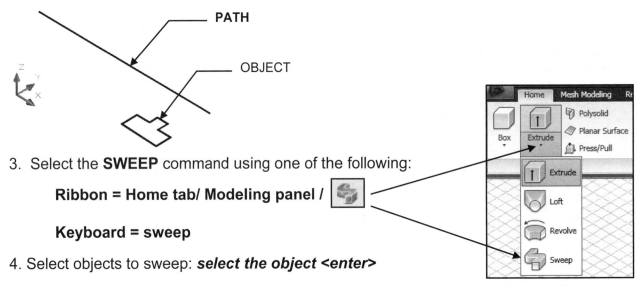

3. Select the **SWEEP** command using one of the following:

    Ribbon = Home tab/ Modeling panel /

    Keyboard = sweep

4. Select objects to sweep: *select the object <enter>*

5. Select objects to sweep: *select more or <enter> to stop*

6. Select sweep path or [Alignment/Base Point/Scale/Twist]: *select the path*

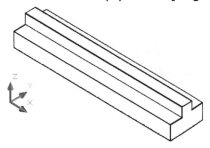

If you would like it to twist, when prompted for the path, select "Twist". Then enter the total angle that you would like it to twist along the entire path. Then select the path.

Example of 180 degree twist from the start of the path to the end.

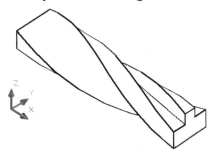

HELIX

The HELIX command allows you to easily construct a wireframe helix or spiral. You will specify the center of the base and top, diameters of the base and top, number of turns, turn height distance and which direction it turns. The default is counterclockwise.

HOW TO USE THE HELIX COMMAND

1. Select the Helix command using one of the following:

   Ribbon = Home tab / Draw panel ▼ /

   Keyboard = helix

Command: _Helix
Number of turns = 3.000   Twist=CCW

1. Specify center point of base: 0,0: *select the location for the center of the base*

2. Specify base radius or [Diameter] <1.000>: *enter base radius <enter>*

3. Specify top radius or [Diameter] <2.000>: *enter top radius <enter>*

4. Specify helix height or [Axis endpoint/Turns/turn Height/tWist] <1.000>: *enter Axis endpoint or select one of the other options.*

Options:

Axis endpoint = center of the top of the Helix.

Turns = How many turns do you want? (500 max) Then you will be prompted for distance between turns. AutoCAD will calculate the height from that information.

Turn Height = Distance between revolutions.

Twist = The default twist is CCW. You may change it to CW.

You may combine the Helix and Sweep commands to create a 3D solid.
First create the Helix. Second draw the cross-section object such as the diameter of a spring. Then use the cross-section as the Sweep object and the Helix as the path.

EXERCISE 19A
REVOLVE

1. Start a **NEW** file and select **Acad3d.dwt.**
2. Draw the 2D shape shown below.
3. Create 2 additional copies (3 total)
4. Revolve in the X axis, Y axis and around an object, **as shown on the next page**.
5. Save as **EX-19A**

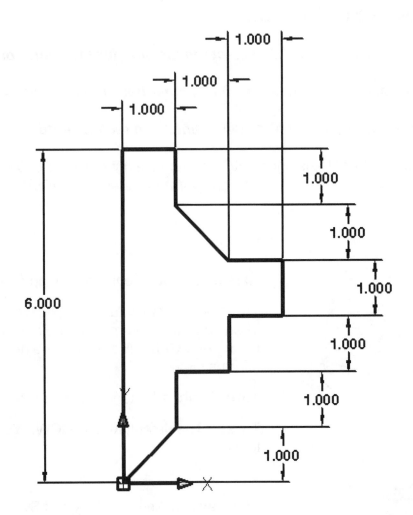

ORIGINAL SHAPE

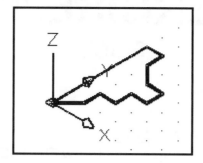

REVOLVE 180 DEGREES AROUND THE X AXIS

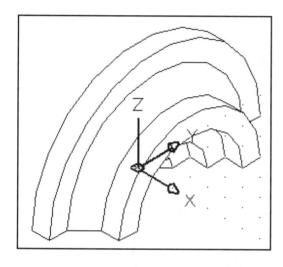

REVOLVE –90 DEGREES AROUND THE Y AXIS

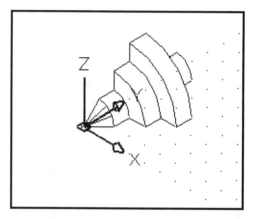

Remember, the point of "Y" is in your chest and you revolve the XZ clockwise (negative).

REVOLVE 360 DEGREES AROUND AN OBJECT

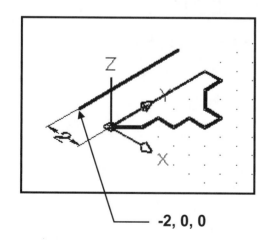

— -2, 0, 0

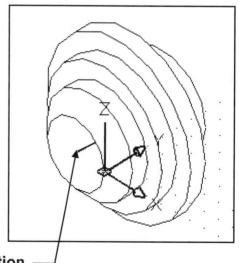

Axis of Rotation —

EXERCISE 19B
SLICE

1. Start a **NEW** file and select **Acad3d.dwt**.
2. Draw the 3D solid shown in the upper right view **FIGURE 1**.
   (The other 3 views are for dimensional information only)
3. Save as **EX-19B1 (This drawing will be used in EX-19C and EX-19D also.)**
4. **SLICE** the solid as shown in **FIGURE 2**.
5. Save as **EX-19B2**

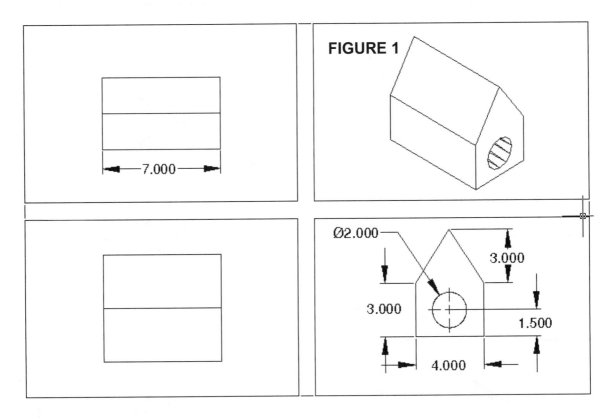

FIGURE 1

Ø2.000

3.000

3.000

1.500

4.000

FIGURE 2

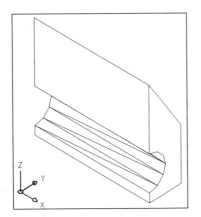

EXERCISE 19C
SECTION

1. Open **EX-19B1 (from the previous exercise)**
2. Using the Section command create a **SECTION** as shown below. (Refer to page 19-4)
3. Save as **EX-19C**

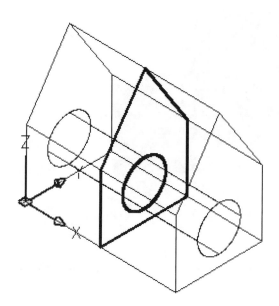

EXERCISE 19D
CREATE 2D and 3D SECTION

1. Open **EX-19B1 (from exercise 19B)**
2. Using **Section Plane** command create a 2D Section as shown below. (Fig. 1)
3. Using **Section Plane** command create a 3D Section as shown below. (Fig. 2)
4. Save as **EX-19D**

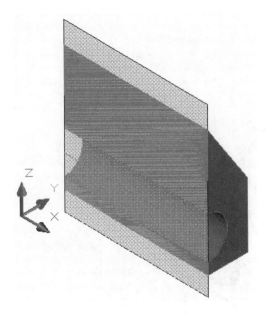

Section Plane location is in the center

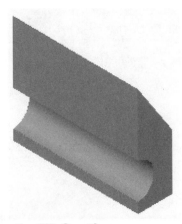

Fig. 1   2D Section   **Fig. 2   3D Section**

EXERCISE 19E
SWEEP

1. Start a **NEW** file and select **<u>Acad3d.dwt.</u>**
2. Draw the 2D shape shown below and the path on the XY plane.

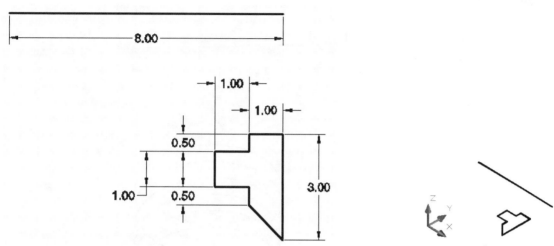

3. Make a Region of the shape.

4. Create a copy of both the Region and the Line.

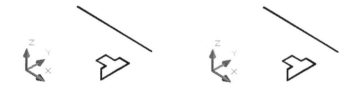

5. Select the layer where you want the sweep placed.

6. First do a straight Sweep. Then do a Sweep with a twist of 180 degrees.
   (Refer to page 19-8)

7. Save as **EX-19E**

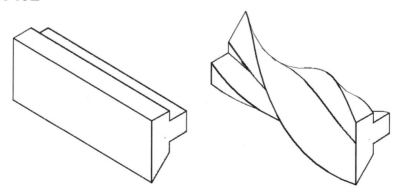

STRAIGHT SWEEP   **TWIST 180 DEGREES SWEEP**

EXERCISE 19F
HELIX

1. Start a **NEW** file and select <u>Acad3d.**dwt.**</u>
2. Draw the 2 HELIX shown below.
3. Save as: **EX-19F**

Number of turns = 10
Twist = CCW
Base radius = 3.000
Top radius = 3.000
Axis end point = 8

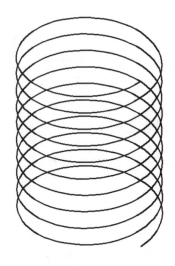

Number of turns = 5   Twist = CCW
Base radius = 6
Top radius = 2
Turns = 5
Turn Height = 1

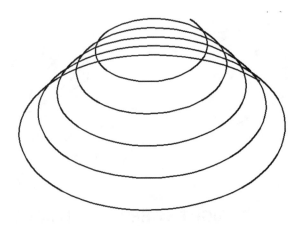

EXERCISE 19G
3D SOLID HELIX

1. **Open EX-19F**
2. Draw the 2 objects shown below each Helix.
3. Sweep the objects (CIRCLE & POLYGON) using the Helix as the Path.
4. Save as: **EX-19G**

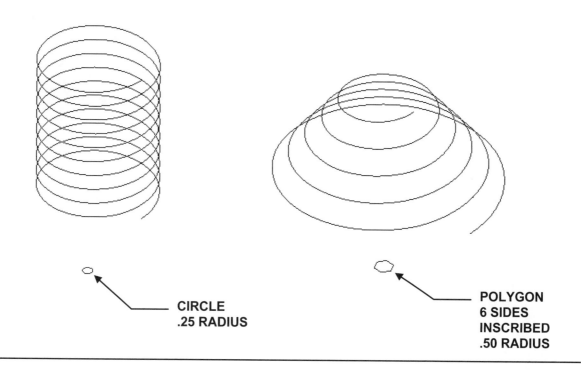

CIRCLE
.25 RADIUS

POLYGON
6 SIDES
INSCRIBED
.50 RADIUS

NOTES:

LEARNING OBJECTIVES

After completing this lesson, you will be able to:

1. Create a multiple view plot setup
2. Shell out a solid model

LESSON 20

PLOTTING MULTIPLE VIEWS

AutoCAD has many commands to create a multiview layout, such as SOLVIEW, SOLDRAW, SOFPROF and DXB plot. You may consider researching these, but in this lesson I will show you 4 quick and simple steps to develop a multiview layout to use for plotting.

STEP 1. SELECTING THE PLOTTER / PRINTER AND PAPER SIZE.

1. Open a solid model drawing. (Example: 2010-3D demo.dwg)
2. Display the model as **Parallel** (See page 13-7)
3. Select an unused Layout tab.
4. **Erase any viewports**
5. Right click on the Layout tab and select **Page Setup Manager**.
6. Select **Modify**
7. Select "your printer", "Plot Style: Monochrome" and " Paper Size-Letter". and OK.
8. Select **Close**
9. Rename the Layout tab to "**Multiview**".

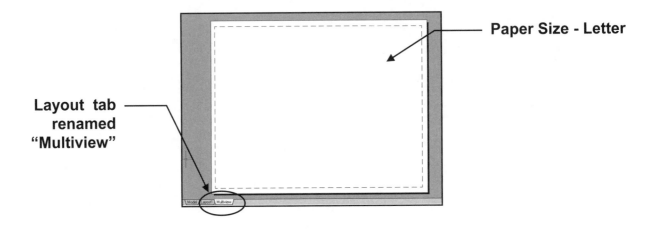

Paper Size - Letter

Layout tab
renamed
"**Multiview**"

STEP 2. CREATE VIEWPORTS
1. Select the **Viewport** layer.
2. Type **vports** <enter>
   a. Select **FOUR EQUAL**
   b. Set **Viewport Spacing to: .50**
   c. Set **Setup to: 3D**
   d. Select **OK**
   e. Specify first corner or [Fit] <Fit>: **<enter>**

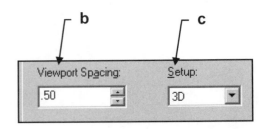

b — Viewport Spacing: .50

c — Setup: 3D

The 3D Model will be displayed within 4 viewports. Each display is a different view of the model.

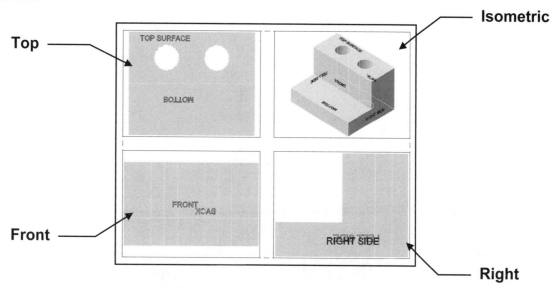

Top

Isometric

Front

Right

STEP 3. ADJUST VIEWPORT SCALE, PAN and LOCK
1. Adjust the scale of each Viewport to **1: 2**.
2. **LOCK** each Viewport.

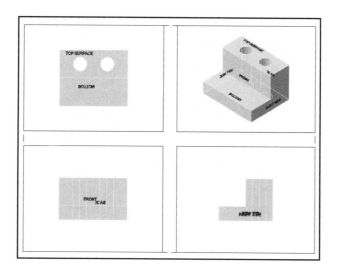

STEP 4. PLOT
1. Select **PLOT** command
2. Select **Printer / Plotter** = yours
3. Select **Paper Size** = Letter
4. Select **Plot Area** = Layout
5. Select **Plot Offset** = 0
6. Select **Scale** = 1 : 1
7. Select **Plot Style Table** = Monochrome
8. Select the **Preview** button.
9. If **OK** then plot.

More Visual Styles

In addition to the Visual Styles shown on page 13-11 you have a few more available.

Select **Visual Styles** tool

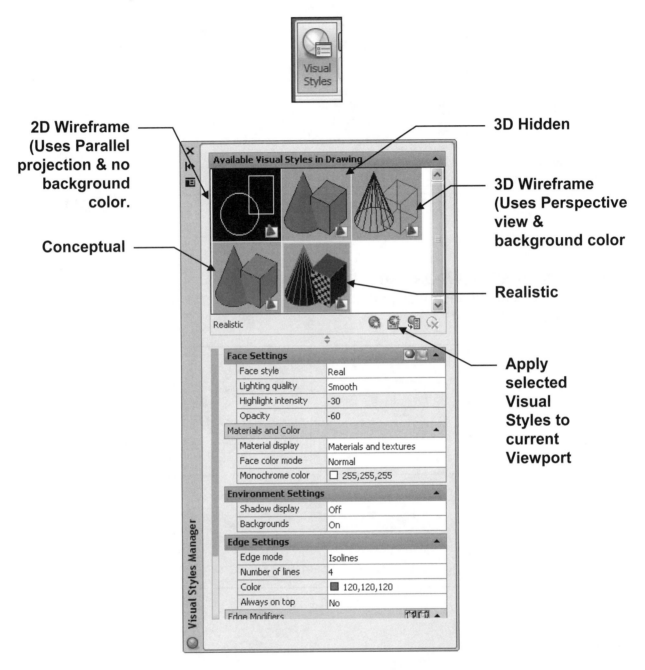

2D Wireframe (Uses Parallel projection & no background color.

3D Hidden

3D Wireframe (Uses Perspective view & background color

Conceptual

Realistic

Apply selected Visual Styles to current Viewport

You many select any of the Visual Styles and then customize the appearance by changing their properties in the lower area of the dialog box. After you have made the changes to the properties, select the **Apply selected Visual Styles to current Viewport** button.

Experiment with these and refer to the help menu for more explanation for each.

SHELL

The SHELL command allows you to hollow out the insides of a solid leaving a thin wall.

How to Shell a solid object:

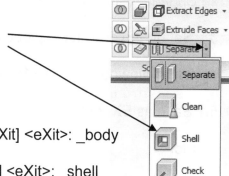

1. Draw a solid object such as a box.

2. Move the UCS origin to the top surface.

3. Select the Shell command using one of the following:

   Ribbon = Home tab / Solid Editing panel /
   Keyboard = Shell

   Command: _solidedit
   Solids editing automatic checking: SOLIDCHECK=1
   Enter a solids editing option [Face/Edge/Body/Undo/eXit] <eXit>: _body
   Enter a body editing option
   [Imprint/seParate solids/Shell/cLean/Check/Undo/eXit] <eXit>: _shell

4. Select a 3D solid: *select the solid object*

5. Remove faces or [Undo/Add/ALL]: *select the top face*

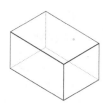

6. Remove faces or [Undo/Add/ALL]: <u>1 face found, 1 removed</u>. *(No visual notification from AutoCAD that you selected a face except this confirmation. Watch for it)*

7. Remove faces or [Undo/Add/ALL]: *select more faces or <enter> to stop*

8. Enter the shell offset distance: *enter the wall thickness <enter>*

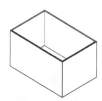

9. Enter a body editing option [Imprint/seParate solids/Shell/cLean/Check/Undo/eXit] <eXit>: *<enter>*

   Solids editing automatic checking: SOLIDCHECK=1

10. Enter a solids editing option [Face/Edge/Body/Undo/eXit] <eXit>: *<enter>*

EXERCISE 20A
PLOT MULTIPLE VIEWS

1. Open **EX-14D**

2. Change display to **Parallel.**

3. Set up the drawing to plot the Top, Front, Right and Isometric views

4. Adjust viewport scale to 1:4 in each viewport and lock.

5. Save as **EX-20A**

6. Plot

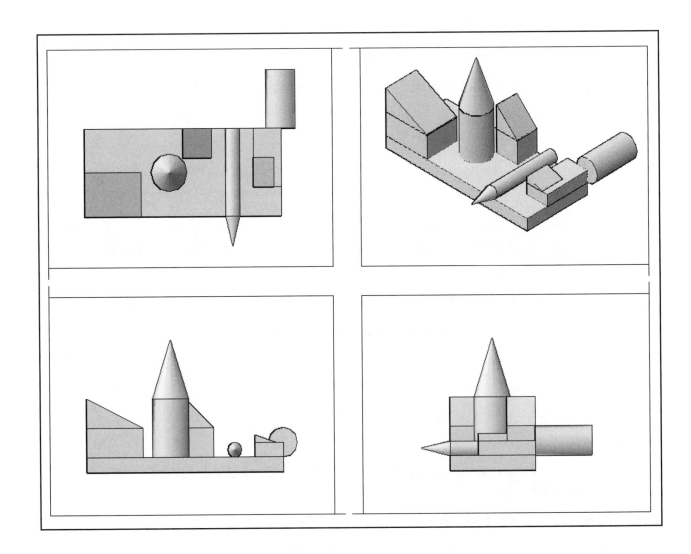

EXERCISE 20B
SHELL

1. **Select File / New and select <u>Acad3d.dwt</u>**

2. Draw the Box shown below. Place a cylinder at each end.
   Box = L = 8   W = 6   Ht = 6
   Cylinder = 2 Radius L = 3

3. Union

4. Shell
   a. Remove the Top Surface
   b. Offset distance = .20

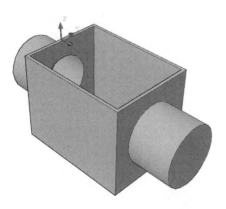

5. Slice through the center line as shown.

Hint for Slice:
1. Move Origin to center line.
2. Enter **ZX** for plane.

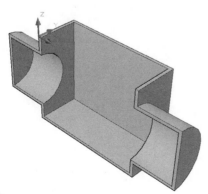

6. Save as **EX-20B**

NOTES:

Welcome to the Architecture Project.

Within this project you will learn one of the many ways to manage a set of drawings required to construct a home. You will also get additional practice drawing, scaling and plotting.

ARCHITECTURE

Architecture Project

In the following pages you will practice how to handle a residential housing construction plan. There are many ways to accomplish this. Consider this just one of them.

The drawings are relatively simple and do not contain all the necessary requirements to build a house. The project is merely designed to walk you through one method of managing the project sheets.

Step 1. Create a **Library** of symbols.

Step 2. Create a new **Master Architecture Border** for 18 X 24 sheet size.

Step 3. Create a site plan

Step 4. Create a Floor Plan

Step 5. Add the Electrical to the Floor Plan

Step 6. Create Elevations

Step 7. Create a Section View

ARCHITECTURAL SYMBOL LIBRARY

When you are using a CAD system, you should make an effort to only draw an object once. If you need to duplicate the object, use a command such as: Copy, Array, Mirror or Block. This will make drawing with CAD more efficient.

In the following exercise, you will create a file full of architectural symbols that you will use often when creating an architectural drawing. You will create them once and then merely drag and drop them, from the DesignCenter, when needed. This will save you many hours in the future.

Save this library file as **Library** so it will be easy to find when using the DesignCenter or create a Library Palette.

1. Open **My Feet-Inches Setup**

2. Select the **Model** tab.

3. Draw each of the Symbol objects, shown on the following pages, actual size.
   Do not scale them. Do not dimension. **Do not make them Annotative**

4. Create an individual Block for each one using the **BLOCK** command.
   Use the number as the name. The actual name is too long.

5. Save this drawing as: **Library**

SYMBOLS

INSTRUCTIONS

1

Ø8"

2"

EXTEND 2"
BEYOND
CIRCLE

1'

DUPLEX CONVENIENCE OUTLET

LAYER = ELECTRICAL

2

GFI

GROUNDED DUPLEX OUTLET

LAYER = ELECTRICAL
GFI TEXT HEIGHT= 3"
POSITION GFI TEXT APPROXIMATELY AS SHOWN
USE A NON-ANNOTATIVE TEXT STYLE

3

GFI

WP

GROUNDED WEATHER PROOF OUTLET

4

220V

220 VOLT OUTLET

5

8"

4"

WALL MOUNTED FIXTURE W/INCANDESCENT LAMP

DIMENSIONS NOT SHOWN ARE THE SAME AS SYMBOL
NUMBER 1.

LAYER = ELECTRICAL

6

2"

CEILING MOUNTED FIXTURE W/FLUORESCENT LAMP

7

SUSPENDED FIXTURE W/INCANDESCENT LAMP

8

2"

RECESSED FIXTURE W/INCANDESCENT LAMP

ARCH-4

SYMBOLS	INSTRUCTIONS

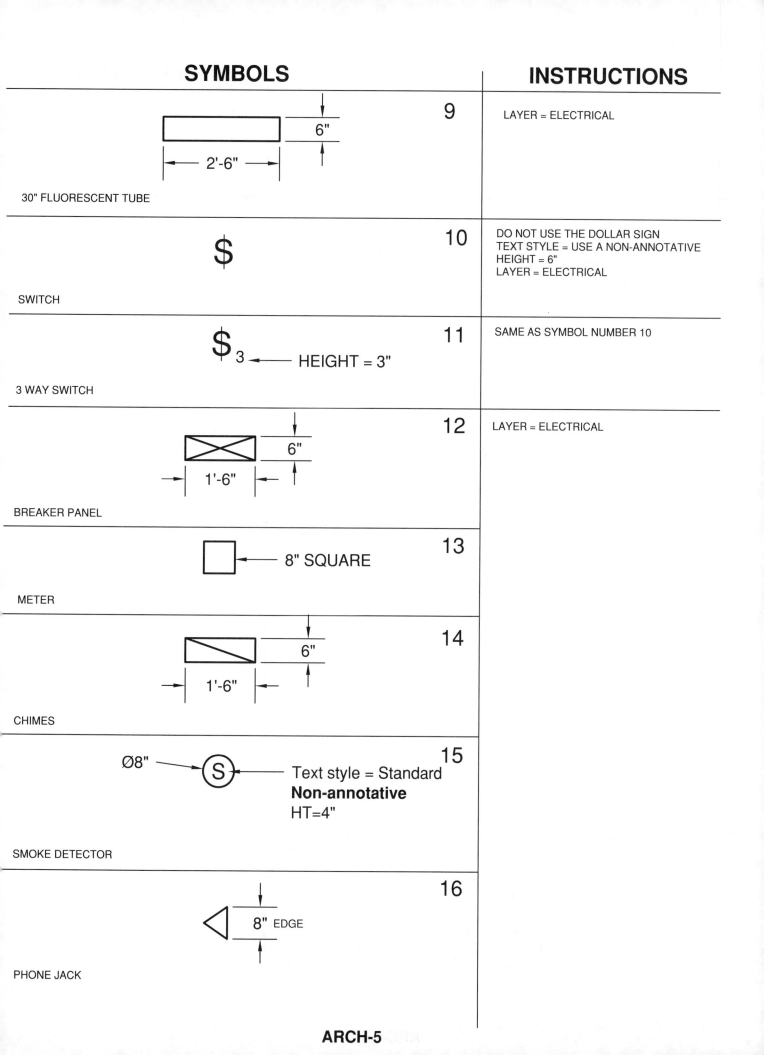

SYMBOLS

9

6"

2'-6"

30" FLUORESCENT TUBE

10

$

SWITCH

11

$3 ← HEIGHT = 3"

3 WAY SWITCH

12

6"

1'-6"

BREAKER PANEL

13

8" SQUARE

METER

14

6"

1'-6"

CHIMES

15

Ø8" → (S) ← Text style = Standard
Non-annotative
HT=4"

SMOKE DETECTOR

16

8" EDGE

PHONE JACK

INSTRUCTIONS

9 LAYER = ELECTRICAL

10 DO NOT USE THE DOLLAR SIGN
TEXT STYLE = USE A NON-ANNOTATIVE
HEIGHT = 6"
LAYER = ELECTRICAL

11 SAME AS SYMBOL NUMBER 10

12 LAYER = ELECTRICAL

ARCH-5

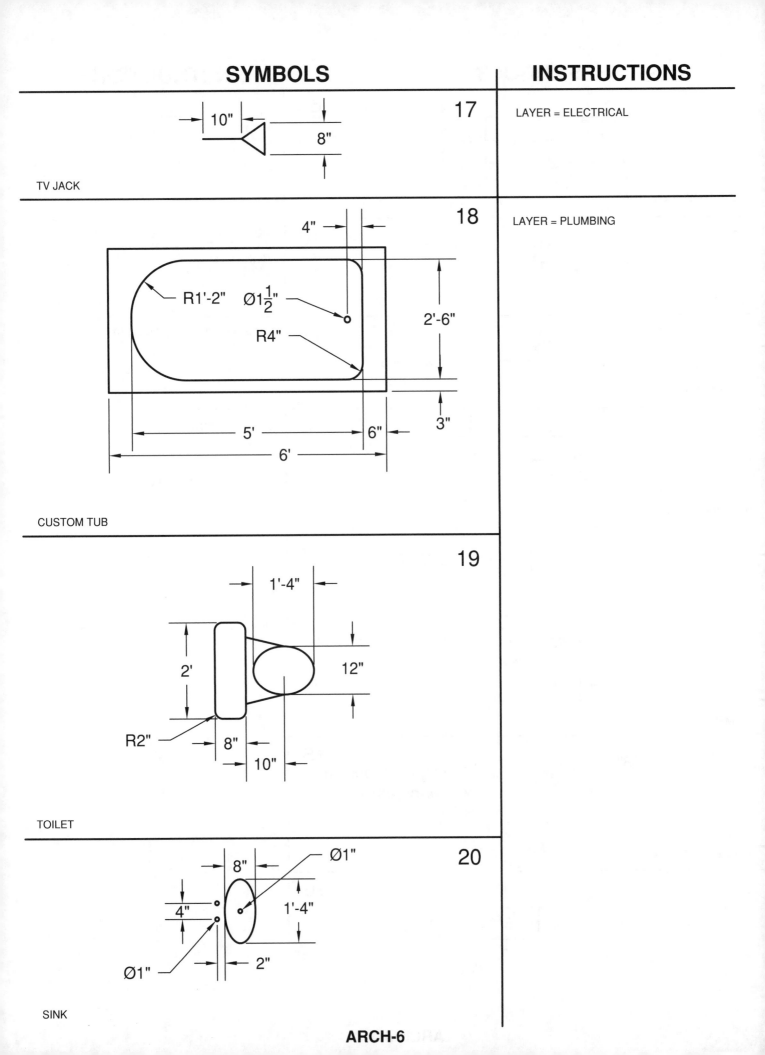

17

LAYER = ELECTRICAL

10"

8"

TV JACK

18

LAYER = PLUMBING

4"

R1'-2"   Ø1½"

R4"

2'-6"

3"

5'   6"

6'

CUSTOM TUB

19

1'-4"

2'

12"

R2"   8"

10"

TOILET

20

8"   Ø1"

4"

1'-4"

Ø1"

2"

SINK

SYMBOLS

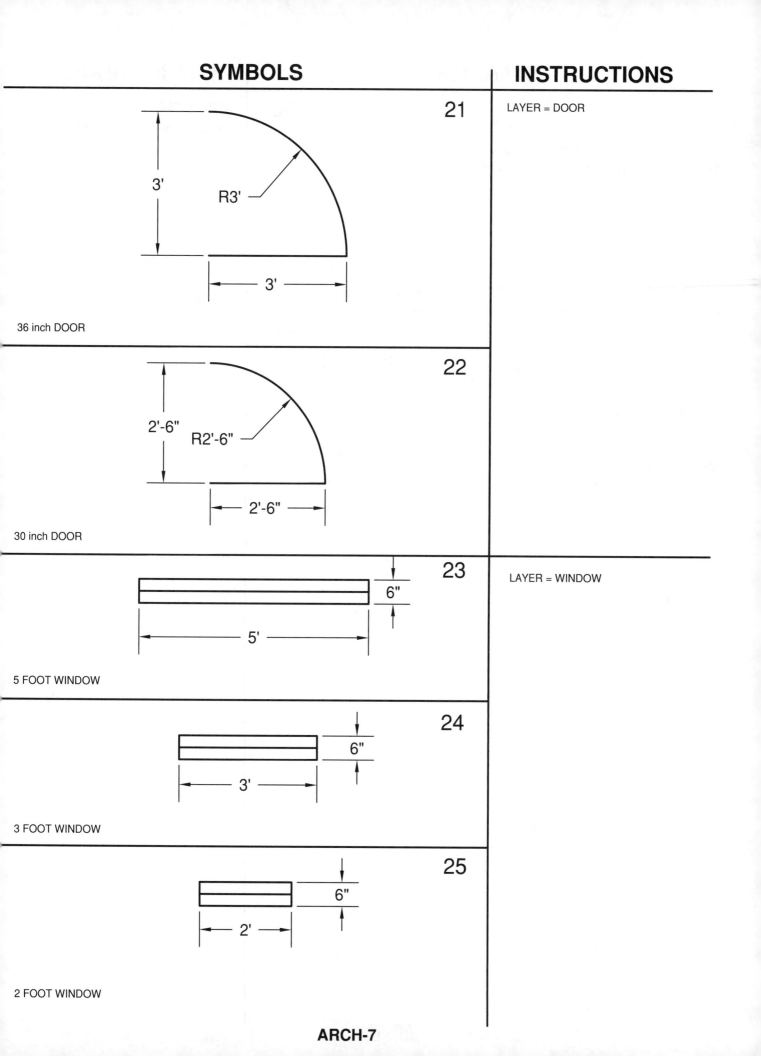

21

3'

R3'

3'

36 inch DOOR

22

2'-6"

R2'-6"

2'-6"

30 inch DOOR

23

6"

5'

5 FOOT WINDOW

24

6"

3'

3 FOOT WINDOW

25

6"

2'

2 FOOT WINDOW

INSTRUCTIONS

LAYER = DOOR

LAYER = WINDOW

ARCH-7

SYMBOLS	INSTRUCTIONS

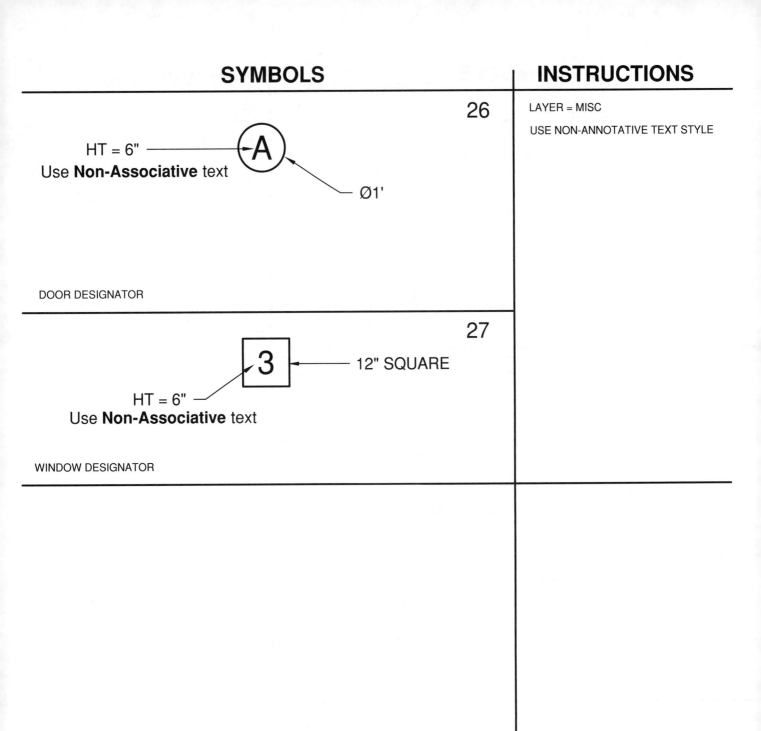

26

HT = 6" — **A**

Use **Non-Associative** text

Ø1'

DOOR DESIGNATOR

27

3 ← 12" SQUARE

HT = 6" —

Use **Non-Associative** text

WINDOW DESIGNATOR

LAYER = MISC

USE NON-ANNOTATIVE TEXT STYLE

Create a New Border

Before you can work on the following exercises you must first create a new Page Setup and Border for a larger sheet of paper. Follow the following steps.

PAGE SETUP

A. Open **My Feet-Inches Setup**.

B. Select a **LAYOUT2** tab.

B. If you do not have these tabs refer to page 2-17

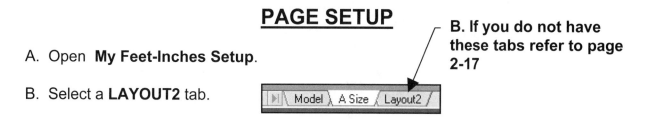

Note: If the "Page Setup Manager" dialog box shown below does not appear automatically, right click on the Layout tab, then select Page Setup Manager.

C. Select the Modify button.

Yours may be different. That's OK for now.

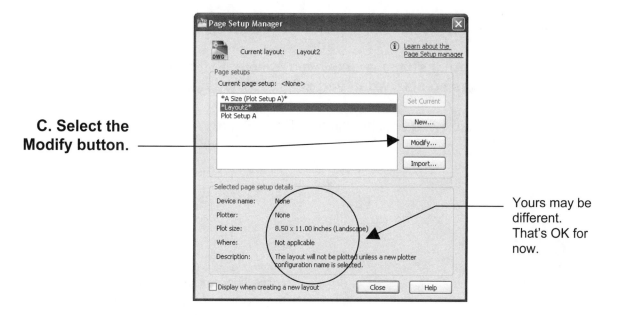

The "Page Setup" dialog should appear.

D. Select DWF6ePlot.pc3

E. Select ARCH (24.00 x 18.00 Inches)

F. Select Layout

G. Plot offset 0

H. Select the OK button.

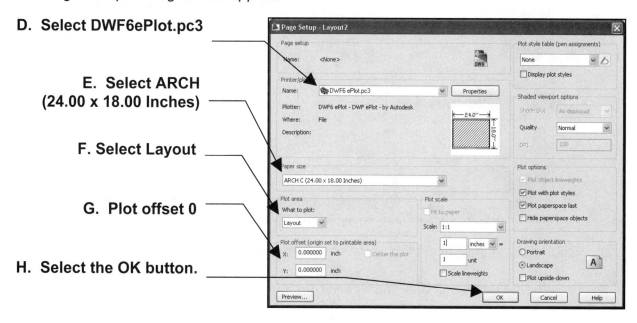

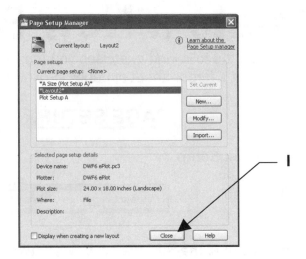

I. Select the **Close** button.

You should now have a sheet of paper displayed on the screen.

Note: If a rectangular shape is displayed, delete it. It is a Viewport that automatically was created when you selected the Layout tab.

This sheet is the size you specified in the "Page Setup". (24 X 18)

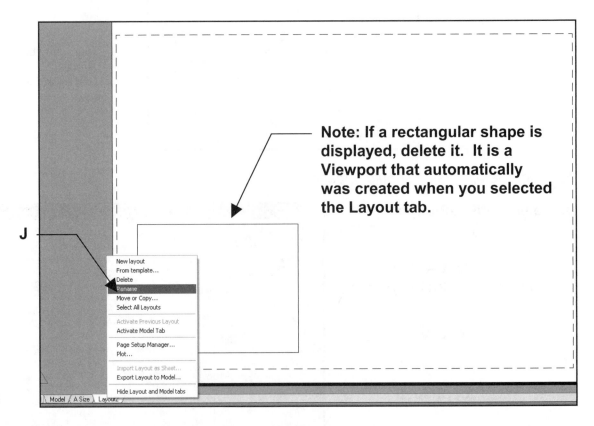

Note: If a rectangular shape is displayed, delete it. It is a Viewport that automatically was created when you selected the Layout tab.

J. Right click on the Layout tab and select **Rename**.

K. Type the new name **C Size Master <enter>**

L. **Draw the Border with title block**, shown below, on the sheet of paper shown on the screen.

M. When you have completed the Border, shown below:
Save as: "Master Architecture Border"

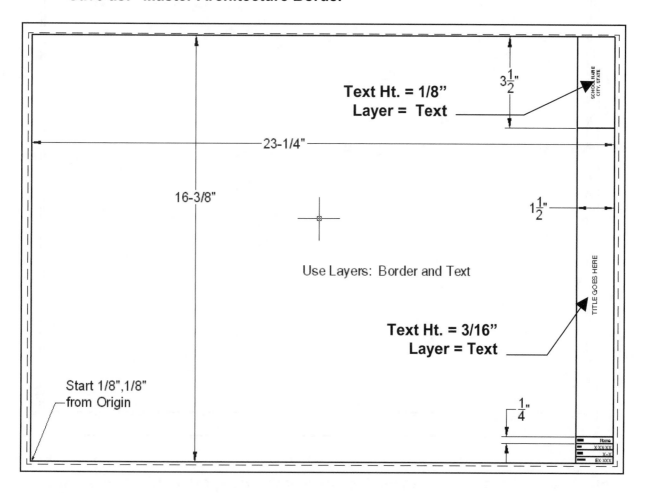

Use Text Style = Text-Classic

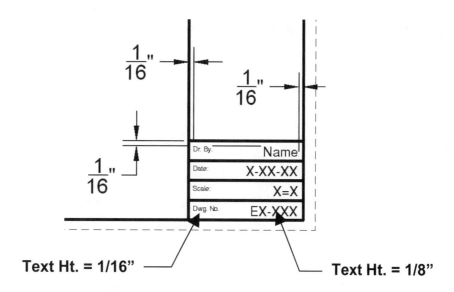

N. **Create a Viewport** approximately as shown below.
   1. Use Layer **Viewport**.

Note: It is not necessary to adjust the scale of the viewport at this time because you will be adjusting the Viewport scale for each exercise.

O. Save again as: **Master Architecture Border**

Continue on to the next page.....

P. Create a new **Plot Page Setup**
   1. Select **Plot**
   2. Change the settings to match settings shown below:

Q

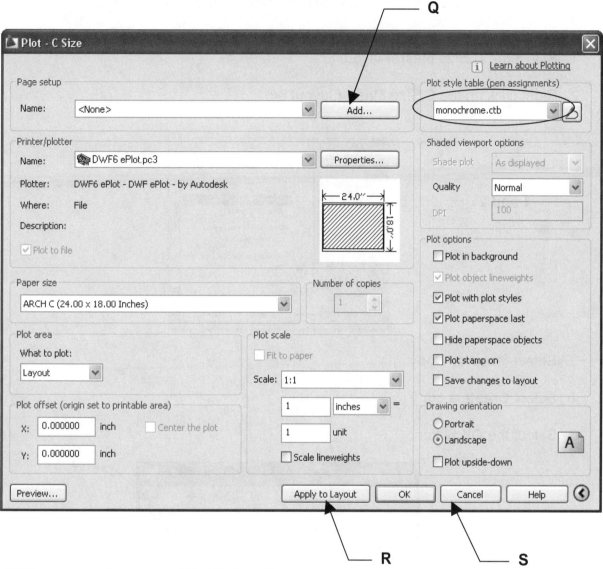

Q. **ADD** a page setup name: **Plot Setup C**

R. **Apply to Layout**

S. Select **Cancel** (This closes the dialog box but your settings are still saved)

T. Save again: **Master Architecture Border**

Continue on to the next page…..

Create additional layouts

Now you are going to learn how to duplicate a layout.

1. Select the **C Size Master** layout tab.

2. Right click on the **C Size Master** layout tab.

1 and 2 ——————

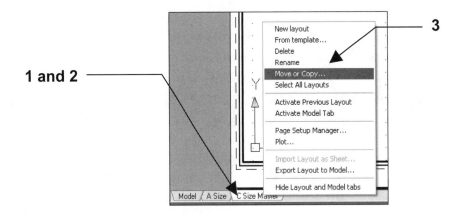

3. Select **Move or Copy…**

4. Select **Create a copy**

5. Select the **OK** button.

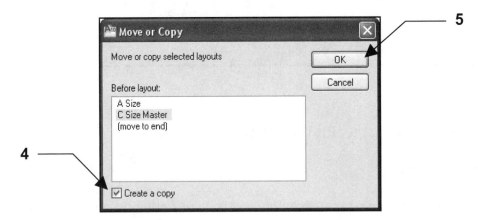

Notice the new layout tab "C Size Master (2)" has been created.

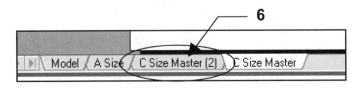

6. Select the **C Size Master (2)** layout tab

Now you should see an exact duplicate of your previously created "C Size Master" layout.   Think about what a time saving process this is.

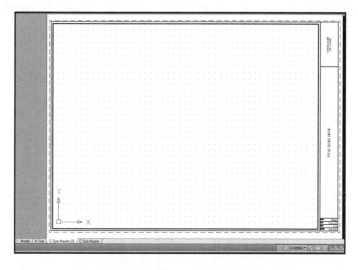

Now you need to make a few changes to the new layout.

7. Double click inside the Viewport until you see the **Model space Origin**. (Make sure you are in Model)

8. Change the Model space Limits: (Type: **Limits**)
    Lower left corner = **0, 0**   Upper Right corner = **170', 130' (feet)**

9. Zoom / All (*If you grids are on you will be able to see what is happening*)

10. Adjust the Viewport Scale to **1/8" = 1'** and **Lock**

11. Change the Title to: **Site Plan**

12. Rename the Layout tab to: **Site Plan**

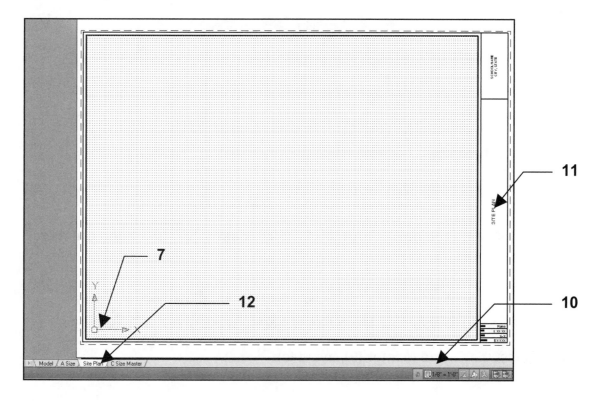

**Now you have a layout tab prepared for a "Site Plan".
The Border is already there and all of the Plot settings are already set.
This process copied all of the settings and paperspace objects into the new layout. This is a big time saver.**

Now you need to do it all again.

13. Select the **C Size Master** layout tab.

14. Right click on **C Size Master** layout tab.

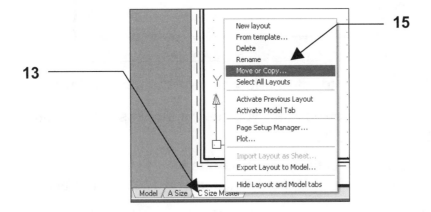

15. Select **Move or Copy…**

16. Select **Create a copy**

17. Select the **OK** button.

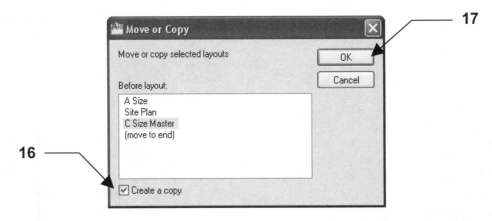

Notice the new layout tab "C Size Master (2)" has been created.

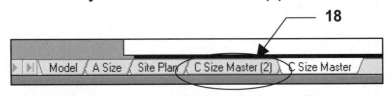

18. Select the **C Size Master (2)** layout tab

You should see another duplicate.

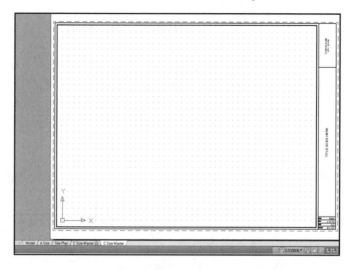

Now you need to make a few changes to the new layout.

19. Double Click inside the Viewport until you see the **Model space Origin**.
    Make sure you are in Model

20. Zoom / All (*If your grids are on you will see what is happening*)

21. Adjust the Viewport Scale to **1/4" = 1'** and **Lock**

22. Change the Title to: **Floor Plan**

23. Rename the Layout tab to: **Floor Plan**

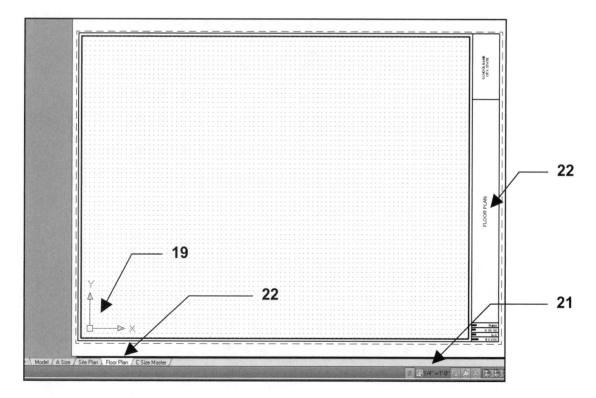

Now you have a layout tab prepared for a "Floor Plan".

Now you need to do it all again.

24. Select the **Floor Plan** layout tab this time.

25. Right click on the **Floor Plan** layout tab.

24 and 25

26

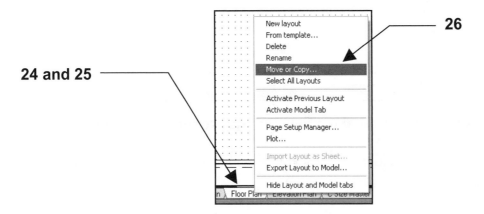

26. Select **Move or Copy**.

27. Select **Create a copy**

28. Select the **OK** button.

28

27

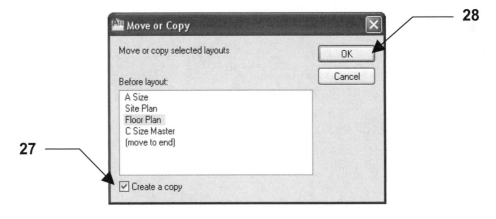

Notice the new layout tab "Floor Plan (2)" has been created.

29

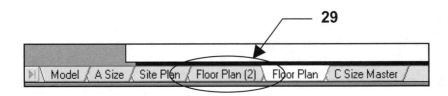

29. Select the **Floor Plan (2)** layout tab

You should see another duplicate.

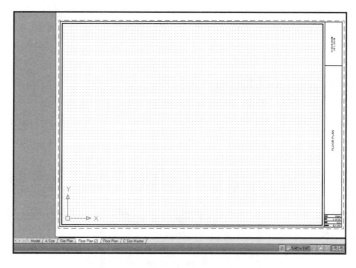

Now you need to make a few changes to the new layout.

30. Change the Title to: **Electrical Plan**

31. Rename the Layout tab to **Electrical Plan**

Note: The viewport scale is the same as the Floor Plan so no need to change.

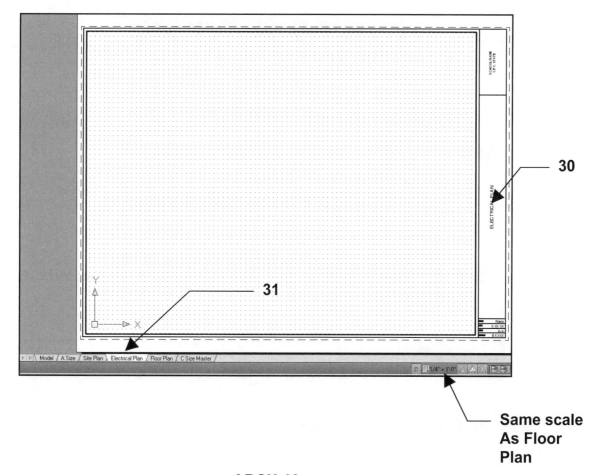

30

31

Same scale
As Floor
Plan

Now you have another layout tab prepared for an "Electrical Plan".

Now you need to do it all again.

32. Select the **Floor Plan** layout tab again.

33. Right click on the **Floor Plan** layout tab.

34

32 and 33

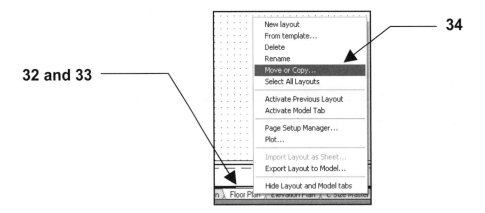

34. Select **Move or Copy…**

35. Select **Create a copy**

36. Select the **OK** button.

36

35

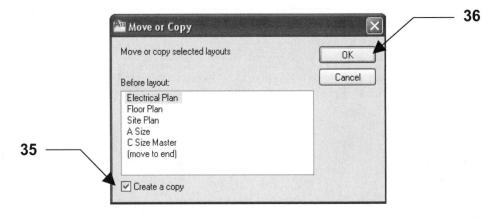

Notice the new layout tab "Floor Plan (2)" has been created.

37

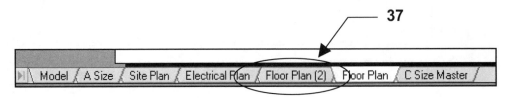

37. Select the **Floor Plan (2)** layout tab

You should see another duplicate.

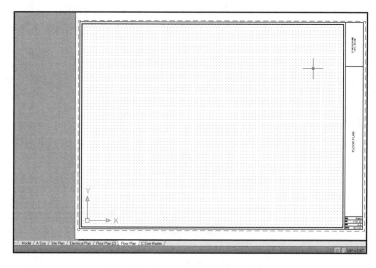

Now you need to make a few changes to the new layout.

38. Unlock the viewport

39. Adjust the Viewport Scale to: 3/8" = 1'

40. Change the Title to: **Elevation Plan**

41. Rename the Layout tab to **Elevation Plan**

Note: The viewport scale is the same as the Floor Plan so no need to change.

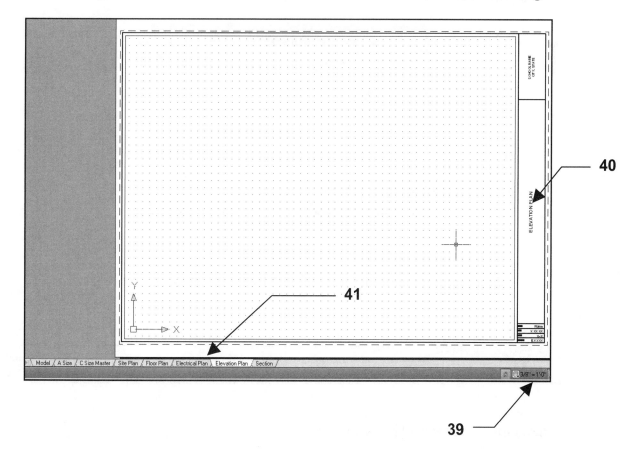

Now you have another layout tab prepared for an "Elevation Plan".

Now you need to do it all <u>one more time</u>.

42. Select the **Floor Plan** layout tab again.

43. Right click on the **Floor Plan** layout tab.

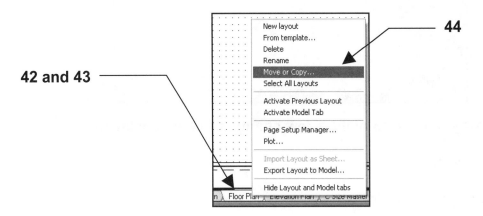

42 and 43   **44**

44. Select **Move or Copy...**

45. Select **Create a copy**

46. Select the **OK** button.

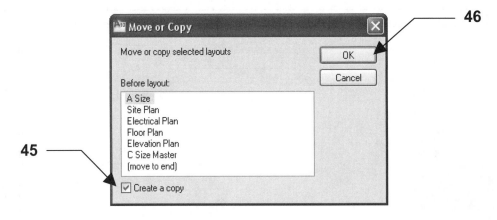

46

45

Notice the new layout tab "Floor Plan (2)" has been created.

47

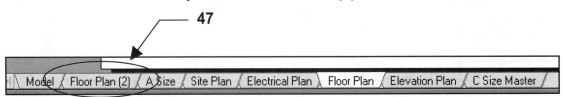

47. Select the **Floor Plan (2)** layout tab

You should see another duplicate.

Now you need to make a few changes to the new layout.

48. Double click inside the Viewport until you see the **Model space Origin**.
    (Make sure you are in Model)

49. Unlock the Viewport, if it is locked.

50. Adjust the Viewport Scale to **1" = 1'** and **Lock**

51. Change the Title to: **Section**

52. Rename the Layout tab to **Section**

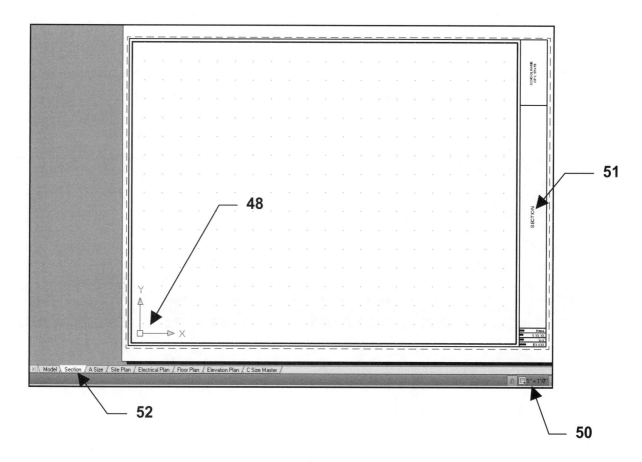

Now you should have 7 Layout tabs and 1 Model tab.

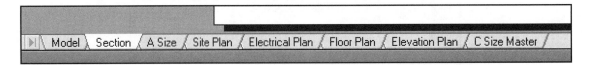

Model \ Section \ A Size \ Site Plan \ Electrical Plan \ Floor Plan \ Elevation Plan \ C Size Master \

Yours may not be in the same order as above, that's OK.

Now I want you to change the display order.

53. Click and drag a tab to the new location and drop it. Try to make yours appear as shown below.

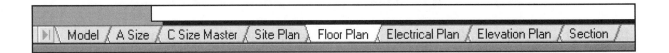

Model \ A Size \ C Size Master \ Site Plan \ Floor Plan \ Electrical Plan \ Elevation Plan \ Section \

54. **Very important:** Save as: **Master Architecture Border**

You now have a Master Border all set up ready to use on the following exercises. If you ever need more layouts, just repeat the steps shown on the previous pages.

Notice if you rest the cursor on any of the Layout tabs a "Thumb size view" of the contents of the Layout will appear.

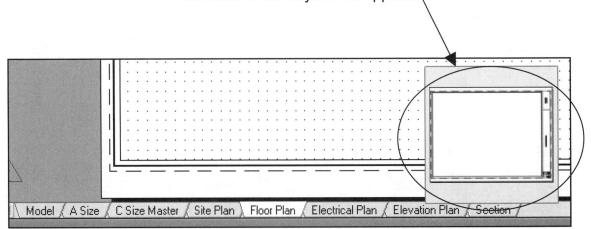

Model \ A Size \ C Size Master \ Site Plan \ Floor Plan \ Electrical Plan \ Elevation Plan \ Section \

EX-ARCH-1

INSTRUCTIONS:

1. Open **Master Architecture Border**
2. Immediately save it as: **ARCH DRAWING SET** (To prevent you from saving on top of Border)
3. Select the **Site Plan** layout tab
4. Draw the **SITE PLAN** shown below.
   Use Layers: Walls, Property Line and Dimension
5. Dimension as shown using Dimension Style: **Dim-Arch**
6. Design a North Symbol in paper space and make it a block (Be creative)
   Use Layer: Symbol
7. **Save** the drawing as: **ARCH DRAWING SET**
8. Plot using Plot page Setup: **Plot Setup C** if you have a large format printer.

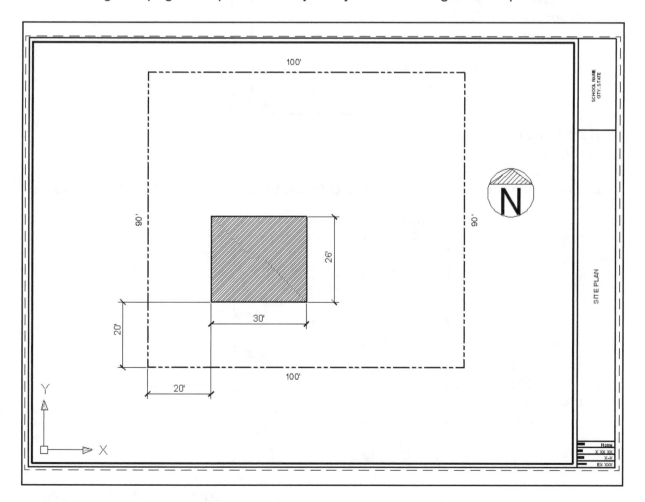

EX-ARCH-2

INSTRUCTIONS:

1. Open **ARCH DRAWING SET**

2. Select the **FLOOR PLAN** layout tab

   Notice the Wall is displayed. It is in Model space and you can see Model space through the viewport. The hatch and dimensions are not displayed because they are a different annotative scale. The Site Plan is 1/8"=1' and the Floor Plan is 1/4"=1'.

Something New

3. Unlock the Viewport and "Pan" the Floor Plan wall to the center of the viewport. You may see part of the "Property Line". We do not want to see the "Property Line" in this layout. So you need to "**Freeze in the Current Viewport**" the Property Line layer.

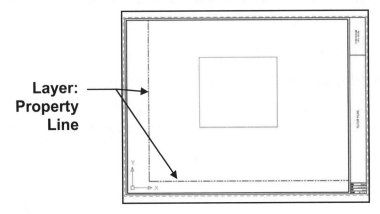

Layer: Property Line

Here is how:
a. Select the Layer ▼
b. Find the "Property Line" layer.
c. Select the "**Freeze or Thaw in the Current Viewport**" icon. (Star changes to blue)

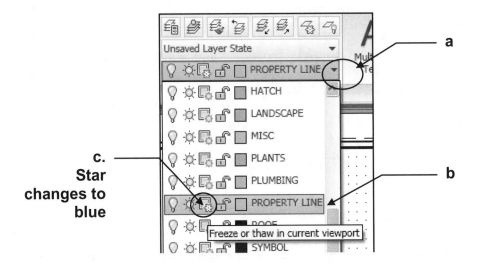

a

c. Star changes to blue

b

Continued on the next page...

The "Property Line" will disappear but **only in the current layout viewport**. If you go back to the "Site Plan" layout you will see that the "Property Line" is still displayed.

Note: Do not use "ON/OFF" or "FREEZE/THAW" in the layer manager. If you do, the layers will disappear in ALL layout viewports.

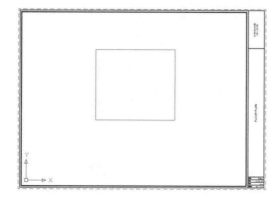

Floor Plan
Layer "Property Line" <u>Frozen</u>
in "Current Viewport" only

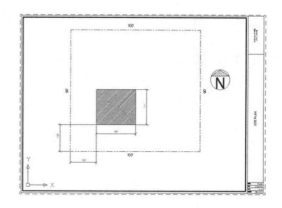

Site Plan
Layer Property Line still displayed
in the "Site Plan" layout

4. Draw the **Floor Plan** shown on the next page.
   Wall width = 6"
   Use Layers: Wall, Door, Window, Plumbing, Dimension and Misc.

5. Dimension as shown using dimension style "Dim-Arch".

6. Draw the Tables in Paperspace on layer Symbol (Size is your option)

7. Create the room labels (Living Room, Bath and Kitchen) using text style "Text-Classic", Height: 1"

8. **Save** the drawing as: **ARCH DRAWING SET**

9. Plot using Plot page Setup: **Plot Setup C** if you have a large format printer.

Continued on the next page...

EX-ARCH-2 continued...

Note: Dimension text height
is shown larger for clarity.
Yours will appear smaller.

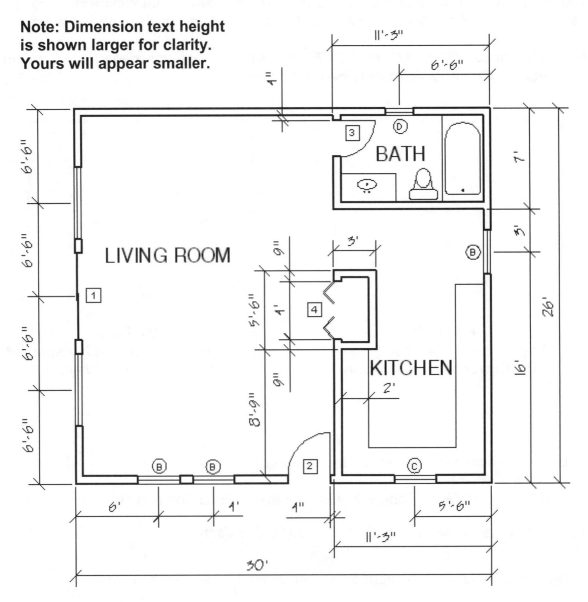

Room labels: Text style: Text-Classic   Height: 1"

Place Schedules in "Paper Space". Size is your choice.

WINDOW SCHEDULE			
SYM	SIZE	TYPE	GLAZING
A	5'-0" X 4'-0"	WOOD FIXED	3/16" SHEET
B	3'-0" X 4'-0"	WOOD FIXED	3/16" SHEET
C	3'-0" X 3'-0"	WOOD FIXED	3/16" SHEET
D	2'-0" X 3'-0"	ALUMINUM SLIDER	3/16" SHEET

DOOR SCHEDULE			
SYM	SIZE	TYPE	MATERIAL
1	6'-0" X 6'-8"	WOOD SLIDER	1/4" POL PL
2	3'-0" X 6'-8"	PANEL	STAIN GRADE
3	2'-8" X 6'-8"	H.C. SLAB	STAIN GRADE
4	4'-0" X 6'-8"	H.C. SLAB	STAIN GRADE

EX-ARCH-2 continued...

Your completed drawing should appear as shown below.

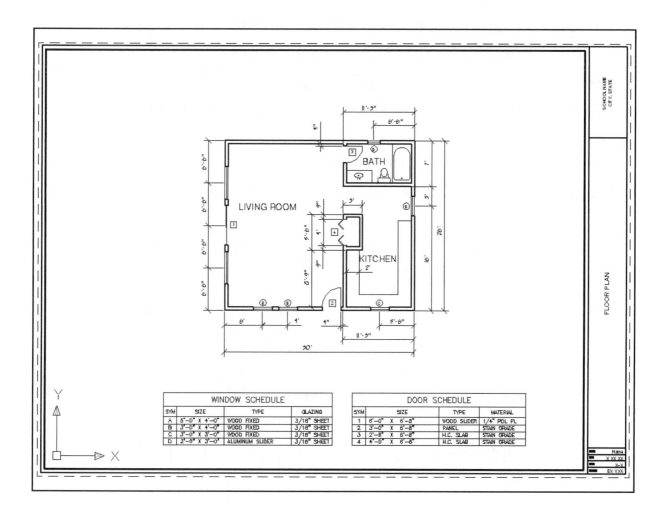

ARCH-29

EX-ARCH-3

INSTRUCTIONS:

1. Open **ARCH DRAWING SET**

2. Select layout tab **Electrical Plan**

3. "**Freeze in <u>Current Viewport</u>**" layers: **Dimension, Misc., Plumbing & Property Line**

4. Use layer "**Wiring**" for the wiring from switches to fixture.

5. Insert the Electrical symbols on <u>layer Electrical</u>.

6. Use Tables to create the Legend of Electrical Symbols on layer Text.

   (Size is your option)

7. **Save** the drawing as: **ARCH DRAWING SET**

8. Plot using Plot page Setup: **Plot Setup C** if you have a large format printer.

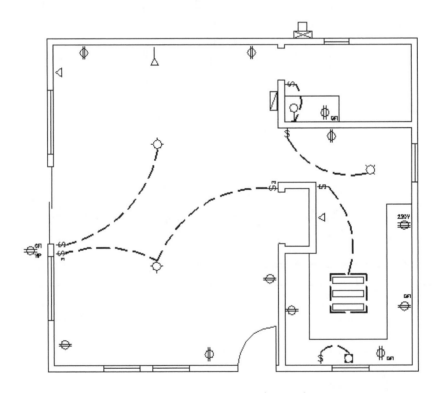

LEGEND OF ELECTRICAL SYMBOLS	
SYM	DESCRIPTION
⊖	DUPLEX CONVENIENCE OUTLET
⊖ᴳᶠᴵ	GROUNDED DUPLEX OUTLET
⊖ᴳᶠᴵ	GROUNDED WEATHER PROOF OUTLET
⊖²²⁰ᵛ	220 VOLT OUTLET
⊢○	WALL MTD. FIXT. W/ INCANDESCENT LAMP
◇	CEILING MTD. FIXT. W/FLOURESCENT LAMP
☼	SUSPENDED FIXT. W/INCANDESCENT LAMP

☼	SUSPENDED FIXT. W/INCANDESCENT LAMP
▭	RECESSED FIXT. W/INCANDESCENT LAMP
▭	30" FLUORESCENT TUBE
$	SWITCH
$₃	3 WAY SWITCH
⊠	BREAKER PANEL
□	METER
▱	CHIMES
Ⓢ	SMOKE DETECTOR
◁	PHONE JACK
─◁	T.V. JACK AND LEAD—IN

EX-ARCH-4

INSTRUCTIONS:

1. Open **ARCH DRAWING SET**

2. Select layout tab **Elevation**

3. "**Freeze in <u>Current Viewport</u>**" layers:

   Cabinets, Dimension, Misc., Plumbing & Property Line

4. Unlock the viewport and adjust the scale to 1/16" = 1'

5. Copy the Floorplan that remained and place the copy on the far right side as shown below.

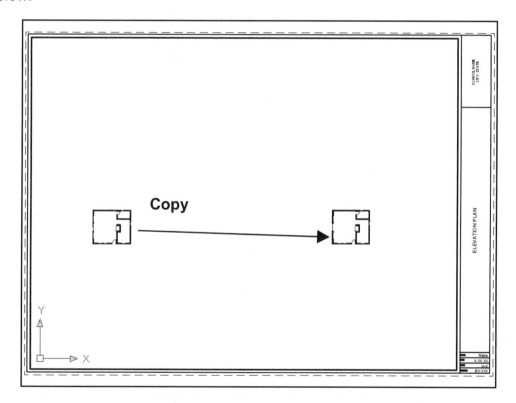

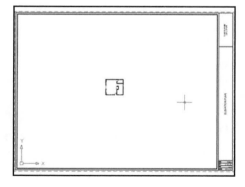

6. Using "Pan" command, pan the copy (located on the right) to the center of the viewport. The original will be out of view.

Continued on the next page...

EX-ARCH-4   continued...

7. Adjust the scale of the Viewport to **3/8" = 1'** and **lock.**

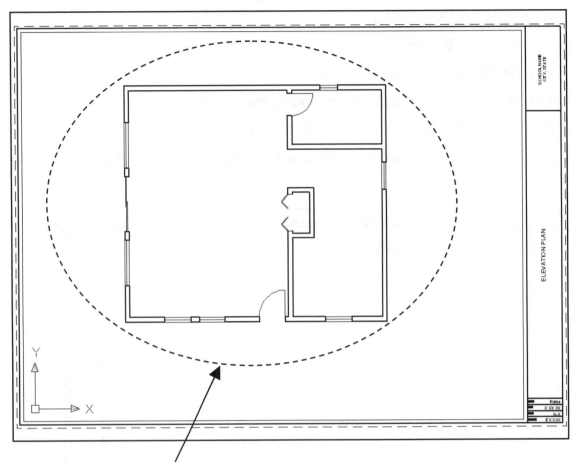

8. Select all of the floor plan and put it on layer "0".

    You can also change the color.

    In the example above I have used color green.

Now using the Floor plan shown above draw the Elevation using the method shown on the next page.

Continued on the next page...

7. Use the Floor plan copy to draw horizontal lines for the Walls, Windows and Door.
    (Refer to the Elevation shown on the next page.)
    Use Object Snap to snap to the Floor Plan to define the horizon location for Walls,
    Windows and Doors.

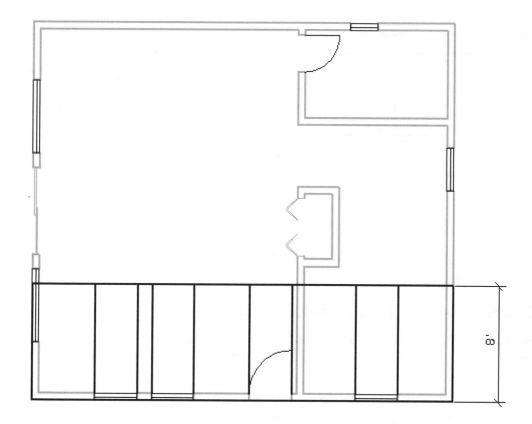

8. Now **"Freeze in current Viewport"** the layer **"0"**
    The Floor plan should have disappeared and only the Elevation Walls remain.

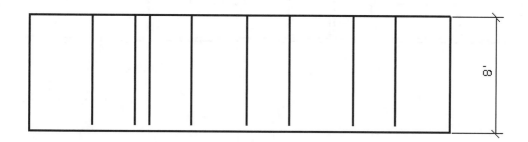

Continued on the next page...

9. Using the "Offset" command define the Vertical heights for the Windows and Door.

   The Top of the Windows and Door is 6'-8".

   The Size of the Windows is defined in the Window Schedule.

   (Refer to the Floor plan layout. page Arch-28)

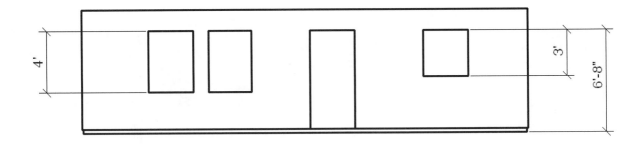

10. The design of the Roof, Windows and Door is your choice. The drawing below is merely an example.

11. Dimension using dimension style: **Dim-Arch**

12. Save as: **ARCH DRAWING SET**

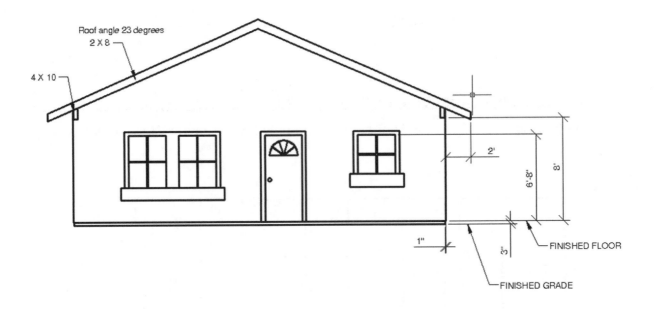

Continued on the next page...

9**If you would like to draw the other 3 views:**

    a. Thaw the "0" layer

    b. Rotate the Floorplan

    c. Draw the horizontal guide lines

    d. When the view is complete, **"Freeze in current Viewport"** to make the Floor-plan disappear.

Note:

You will have to unlock the viewport and use Zoom and pan while you are creating the elevation views. When they are complete, adjust the scale of the viewport and use "pan" to center the drawing. You may have to use 1/4" = 1' to fit all views within the viewport.

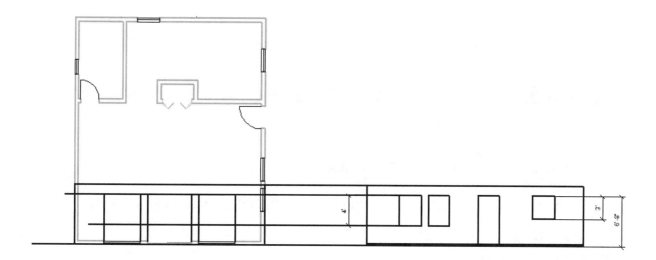

EX-ARCH-5

INSTRUCTIONS:

The following exercise requires that you create a new layer with a new linetype.

1. Open **ARCH DRAWING SET**

2. Select the **SECTION** layout tab

3. If you can see parts of the Floorplan, unlock the viewport and use "Pan" to pan the Floor plan out of view.   Then **lock** the viewport.

4. Create a new layer
   <u>Name</u> = Insulation   <u>Color</u> = magenta   **<u>Linetype = batting</u>**   <u>LWT</u> = default

Now experiment drawing the insulation as follows:
   a. You must be in a layout tab
   b. Select the Layer Insulation
   c. Draw a line

<u>***Note: The insulation linetype is displayed only when you are in a layout tab.***</u>

5. Change the **Linetype scale** as follows:
   a. At the command line type: **LTS <enter>**
   b. Type: **.35 <enter>**
       (This will scale the "batting" linetype to be a little bit smaller than a 2 X 4)

6. Draw the Wall Detail on the next page.
   A. Include the leader call outs.
   b. Use Hatch Pattern "AR-CONC" for footing   Scale = .5
   c. Use User defined for the Finished and Subfloor.
      Angle =45 and 135   Spacing = 3"
   d. Layers and colors, your choice.

7. Save as **EX-ARCH-5**

8. Plot using Plot page Setup: **Plot Setup C** if you have a large format printer.

Continued on the next page...

EX-ARCH-5

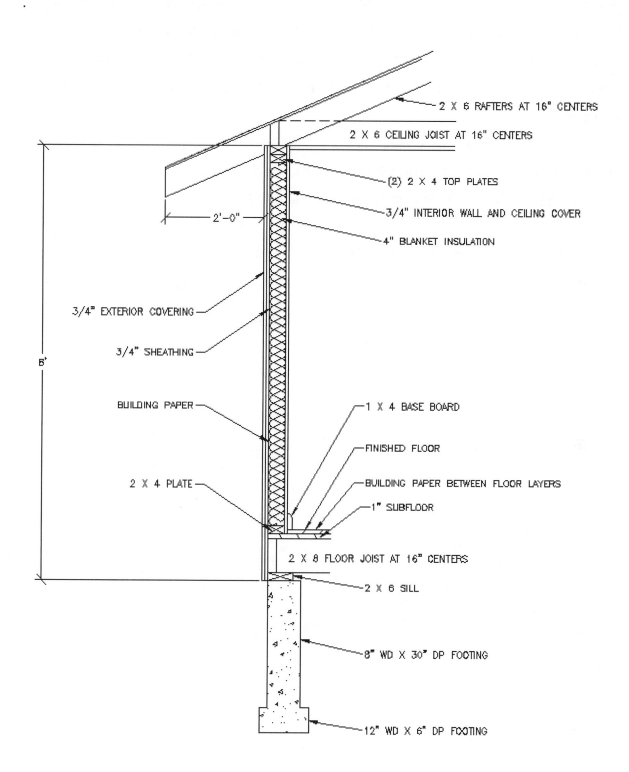

2 X 6 RAFTERS AT 16" CENTERS

2 X 6 CEILING JOIST AT 16" CENTERS

(2) 2 X 4 TOP PLATES

3/4" INTERIOR WALL AND CEILING COVER

4" BLANKET INSULATION

2'-0"

3/4" EXTERIOR COVERING

3/4" SHEATHING

8'

BUILDING PAPER

1 X 4 BASE BOARD

FINISHED FLOOR

BUILDING PAPER BETWEEN FLOOR LAYERS

2 X 4 PLATE

1" SUBFLOOR

2 X 8 FLOOR JOIST AT 16" CENTERS

2 X 6 SILL

8" WD X 30" DP FOOTING

12" WD X 6" DP FOOTING

SUMMARY

.The previous exercises was an example of how you can manage a set of Architectural drawings using multiple layouts and Freezing the layers within the viewports. The advantage to this process is that you never draw anything twice. The disadvantage is that you have to manage your layers. You may even need to create more layers to achieve your goal.

There are other ways to manage your drawings.
Some people prefer to create a separate drawing for each rather than use layout tabs.

Also research **Sheet Sets** in the AutoCAD Help menu.

ELECTRO

MECH

ELECTRO-MECHANICAL SYMBOL LIBRARY

When you are using a CAD system, you should make an effort to ONLY draw an object once. If you have to duplicate the object, use a command such as: Copy, Array, Mirror or Block. Remember, this will make drawing with CAD more efficient.

In the following exercise, you will create a file full of electronic symbols that you consistently use when creating an architectural drawing. You will create them once and then merely drag and drop them, from the DesignCenter, when needed. This will save you many hours in the future.

Save this library file as **Library** so it will be easy to find when using the DesignCenter or create a Library Palette.

1. Start a **NEW** file and select **My Decimal Setup**

2. Add the following layers to your **MY Decimal set up** drawing.

CIRCUIT	GREEN	CONTINUOUS
CIRCUIT2	RED	CONTINUOUS
CORNERMARK	9	CONTINUOUS
DESIGNATOR	BLUE	CONTINUOUS
MISC	CYAN	CONTINUOUS
PADS	GREEN	CONTINUOUS
PCB	WHITE	CONTINUOUS
COMPONENTS	RED	CONTINUOUS

3. Save the **My Decimal Set up**.

4. Select the **Model** tab.

5. Draw each of the Symbol objects, shown on the next page, actual size.
    Do not scale the objects - Do not dimension - Do not make them Annotative.

6. Create an individual Block for each one using the **BLOCK** command. Do not make them Annotative.

7. Save this drawing as: **Library**

SYMBOLS

INSTRUCTIONS

1

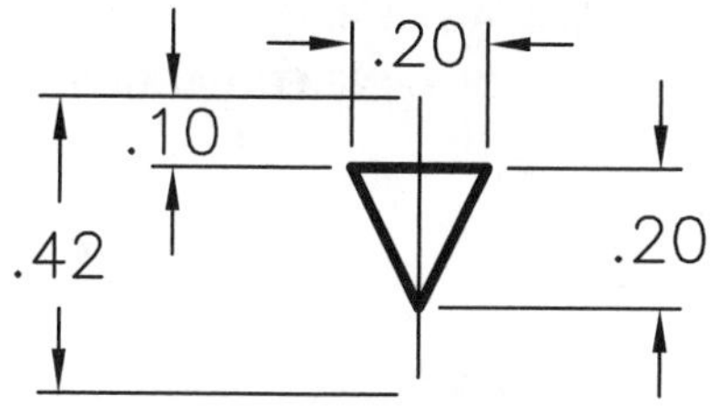

.20

.10

.42

.20

ANTENNA

2

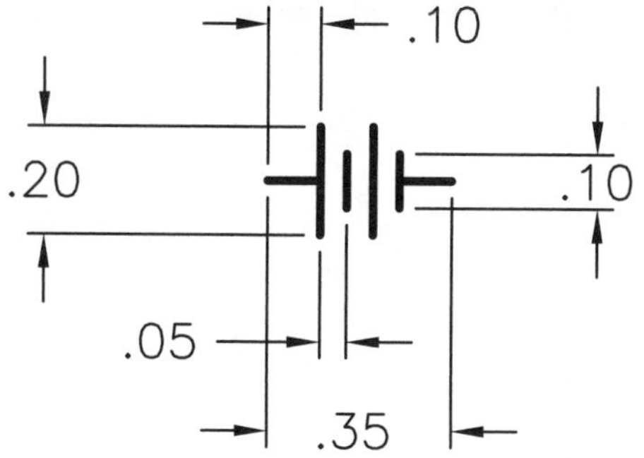

.10

.20

.10

.05

.35

BATTERY

3

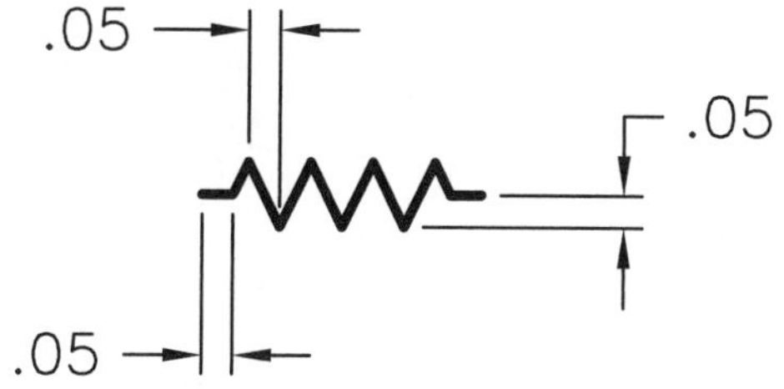

.05

.05

.05

RESISTOR

4

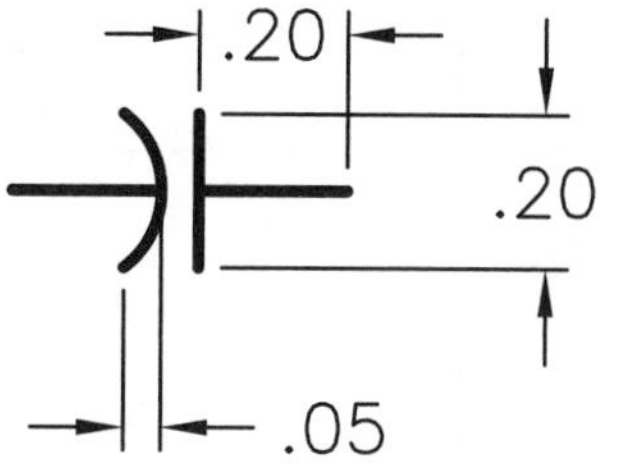

.20

.20

.05

CAPACITOR

5

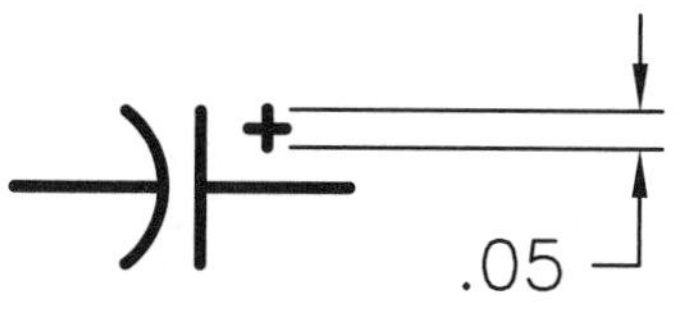

+

.05

CAPACITOR-POLARIZED

ELECT-3

SYMBOLS

INSTRUCTIONS

6

CAPACITOR-VARIABLE

LAYER = COMPONENTS
SAME AS SYMBOL 4
USE POLYLINE FOR ARROW

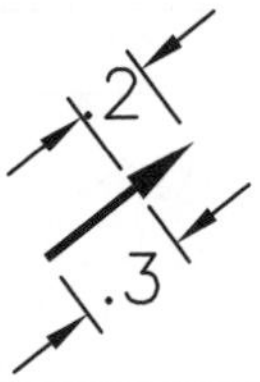

.2

.3

7

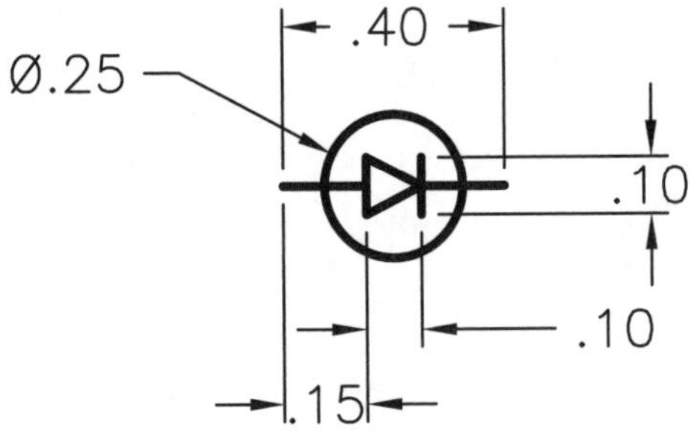

Ø.25

.40

.10

.10

.15

DIODE

LAYER = COMPONENTS

8

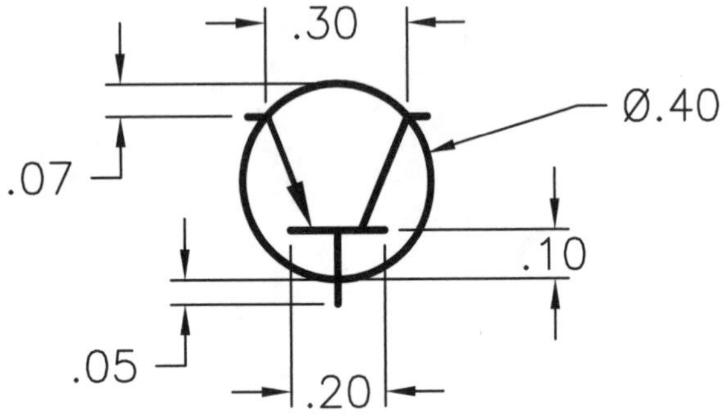

.30

Ø.40

.07

.10

.05

.20

TRANSISTOR-PNP

9

TRANSISTOR-NPN

10

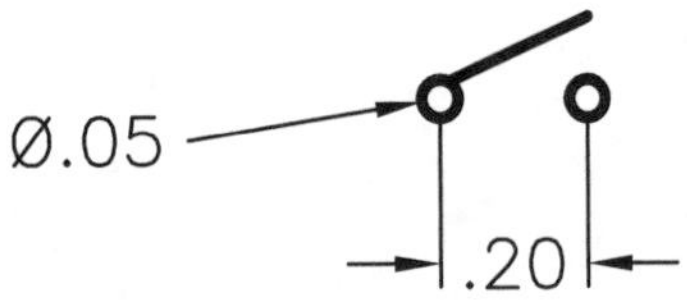

Ø.05

.20

SWITCH

ELECT-4

SYMBOLS

INSTRUCTIONS

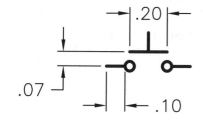

11

LAYER = COMPONENTS

SWITCH-PUSH BUTTON

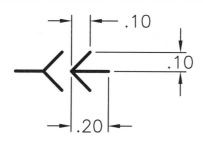

12

PLUG

13

JACK - FEMALE

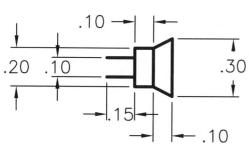

14

SPEAKER

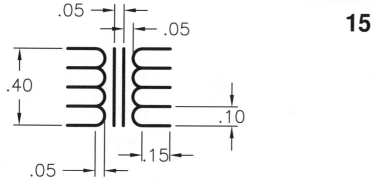

15

TRANSFORMER

ELECT-5

SYMBOLS

INSTRUCTIONS

16

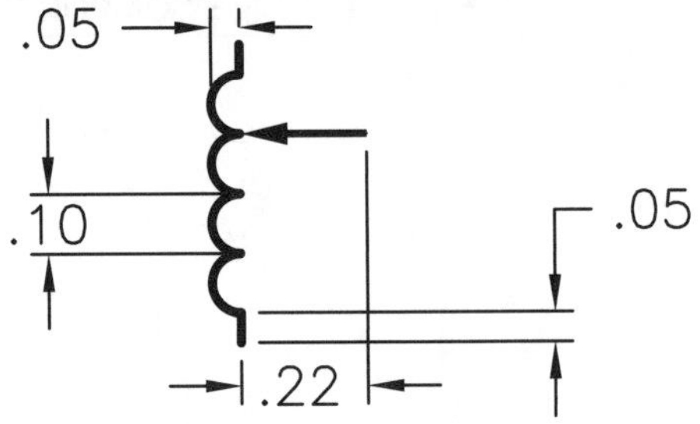

.05

.10

.05

.22

INDUCTOR - VARIABLE

17

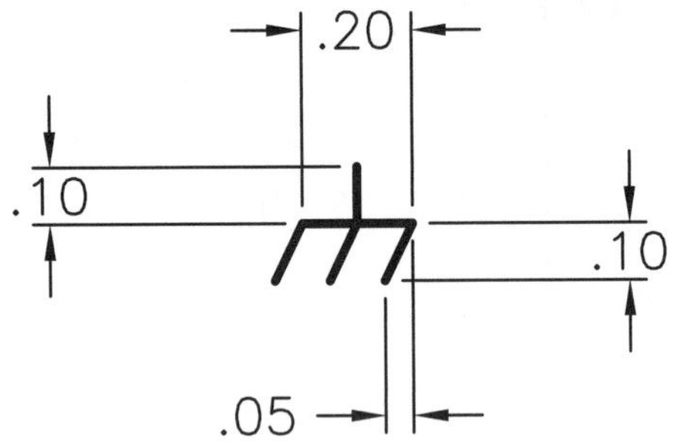

.20

.10

.10

.05

CHASSIS GROUND

18

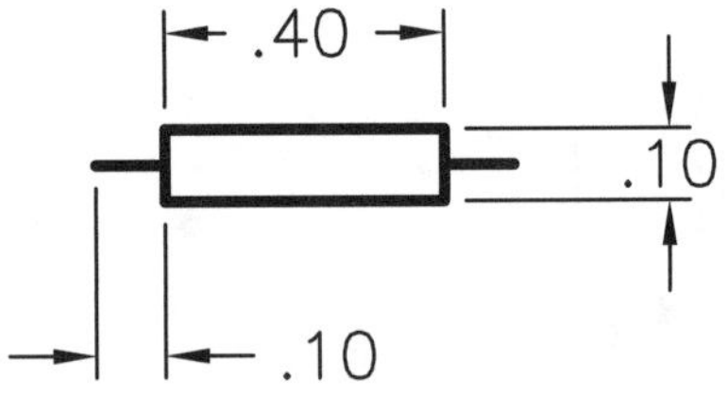

.40

.10

.10

RESISTOR (1/4W)

19

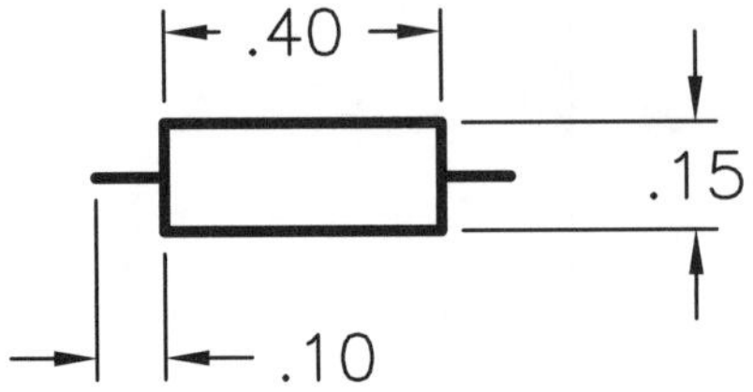

.40

.15

.10

RESISTOR (1/2W)

20

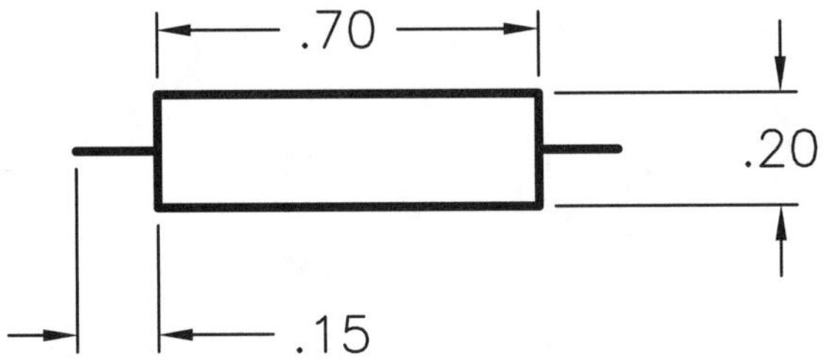

.70

.20

.15

RESISTOR (2W)

LAYER = COMPONENTS

ELECT-6

SYMBOLS

INSTRUCTIONS

21

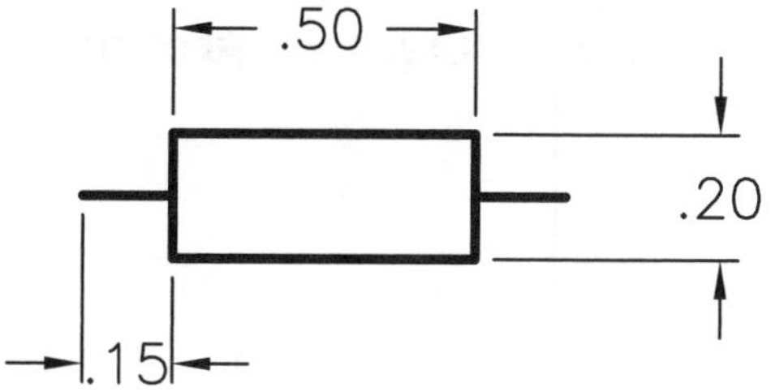

CAPACITOR (22mf)

LAYER = COMPONENTS

22

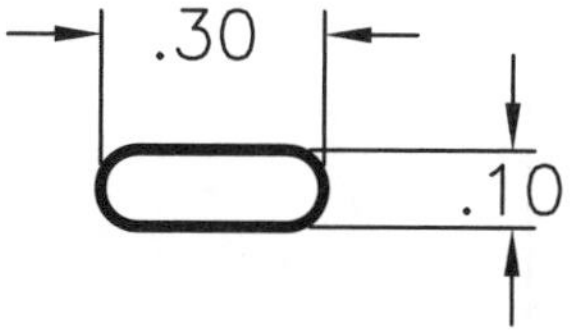

CAPACITOR (.01mf)

23

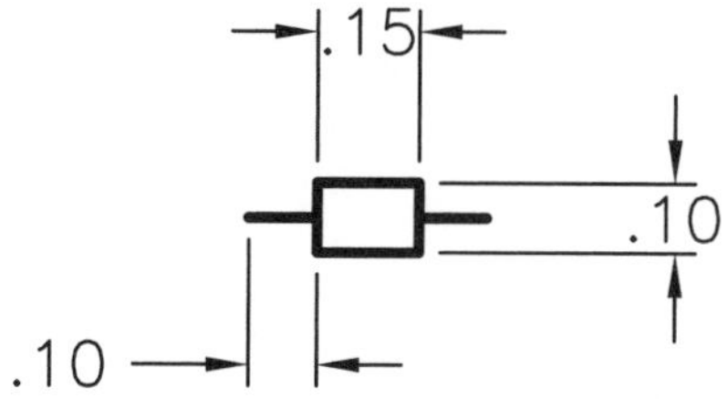

DIODE (IN914)

24

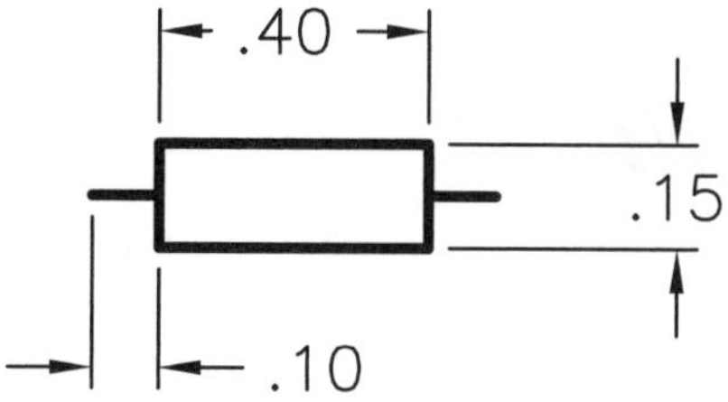

DIODE (IN4001)

25

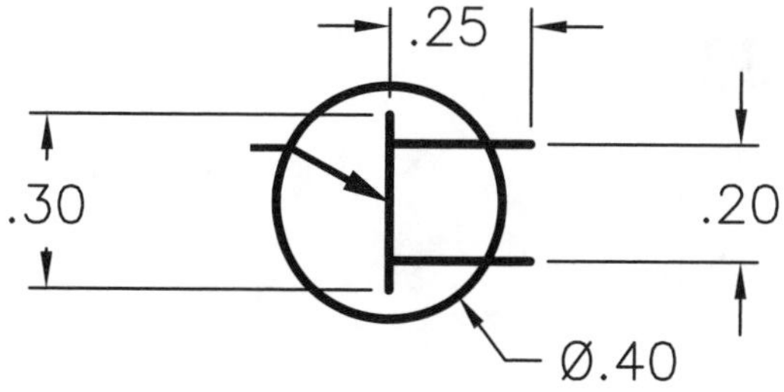

2N2646 UJT (USE FOR SCHEMATIC ONLY)

ELECT-7

SYMBOLS

INSTRUCTIONS

26

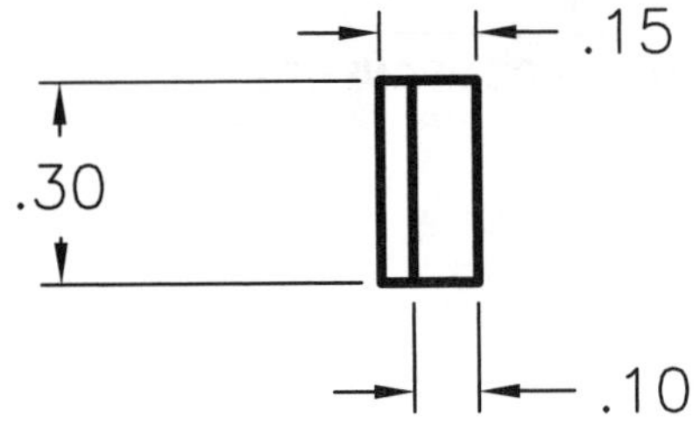

.15

.30

.10

TIP29A

27

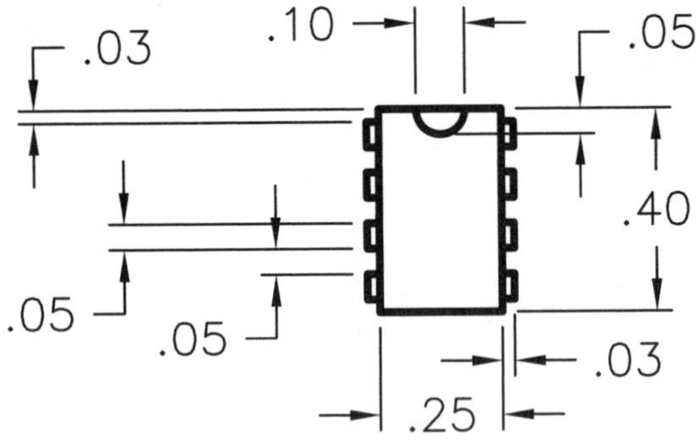

.03

.10

.05

.40

.05

.05

.03

.25

TIMER-NE555

28

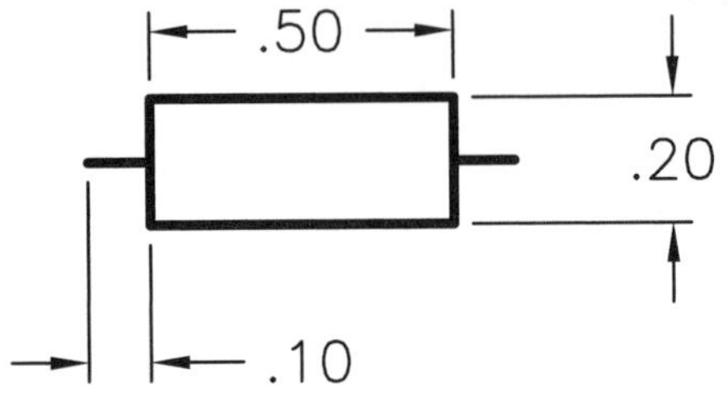

.50

.20

.10

CAPACITOR

29

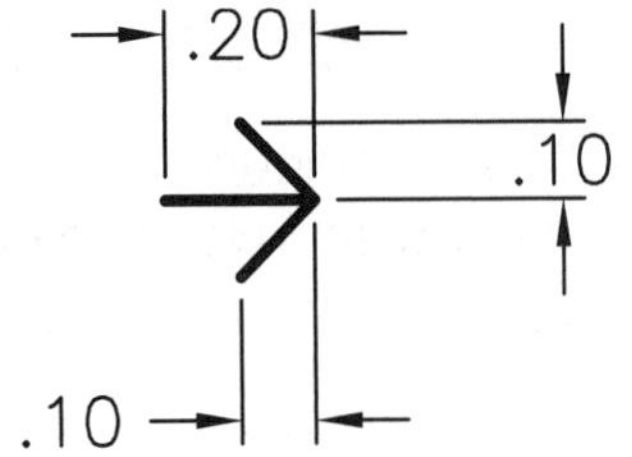

.20

.10

.10

PLUG-MALE

30

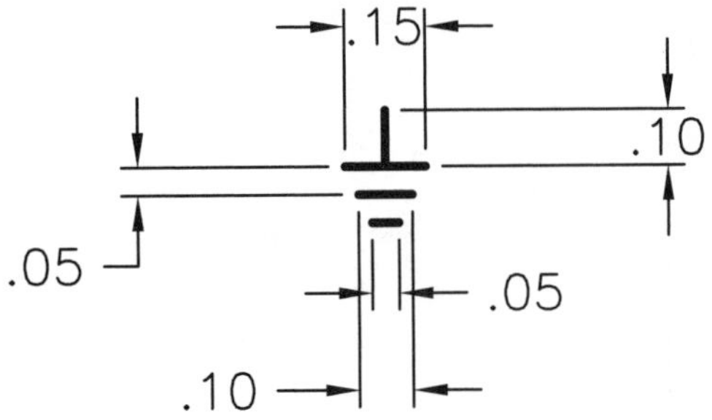

.15

.10

.05

.05

.10

EARTH GROUND

LAYER = COMPONENTS

SYMBOLS		INSTRUCTIONS

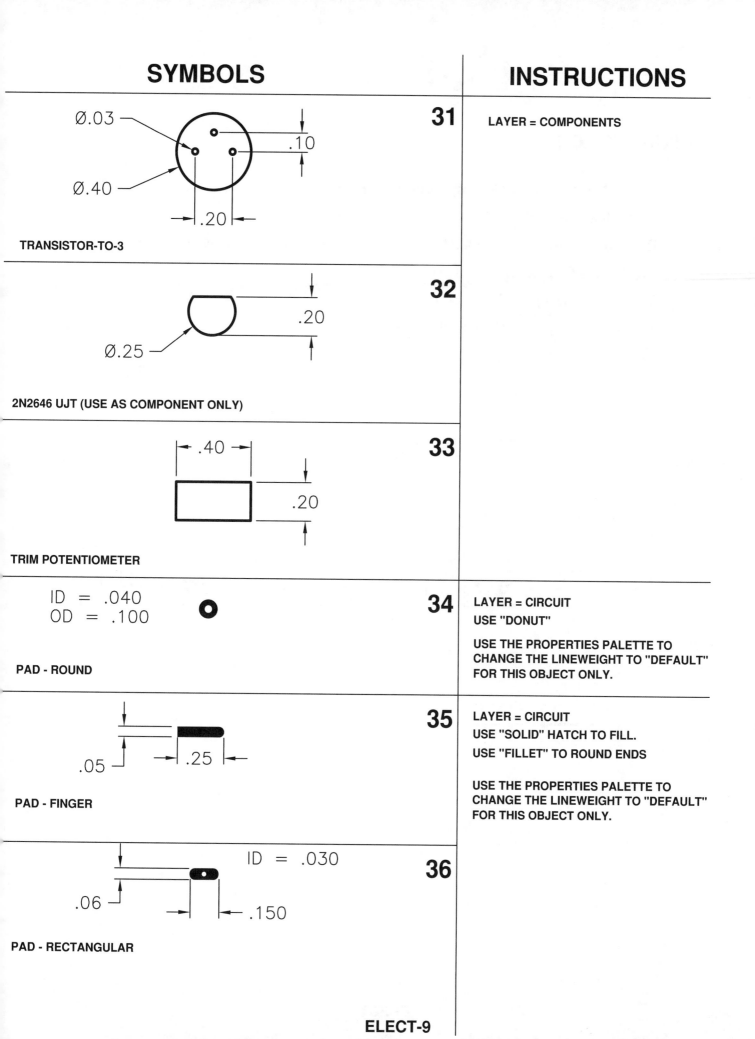

SYMBOLS

31

Ø.03

.10

Ø.40

.20

TRANSISTOR-TO-3

32

.20

Ø.25

2N2646 UJT (USE AS COMPONENT ONLY)

33

.40

.20

TRIM POTENTIOMETER

34

ID = .040
OD = .100

PAD - ROUND

35

.05

.25

PAD - FINGER

36

ID = .030

.06

.150

PAD - RECTANGULAR

INSTRUCTIONS

LAYER = COMPONENTS

LAYER = CIRCUIT
USE "DONUT"

USE THE PROPERTIES PALETTE TO
CHANGE THE LINEWEIGHT TO "DEFAULT"
FOR THIS OBJECT ONLY.

LAYER = CIRCUIT
USE "SOLID" HATCH TO FILL.
USE "FILLET" TO ROUND ENDS

USE THE PROPERTIES PALETTE TO
CHANGE THE LINEWEIGHT TO "DEFAULT"
FOR THIS OBJECT ONLY.

EX-ELECT-1

INSTRUCTIONS:

1. Start a **NEW** file and select **My Decimal Setup**

2. Select the **A Size** layout tab

4. Draw the **BLOCK DIAGRAM** shown below. Size and proportion is your choice.

5. Use Layers **Object line** and **Text**.

6. **Save** the drawing as: **EX-ELECT-1**

7. Plot using Plot page Setup: **Plot Setup A**

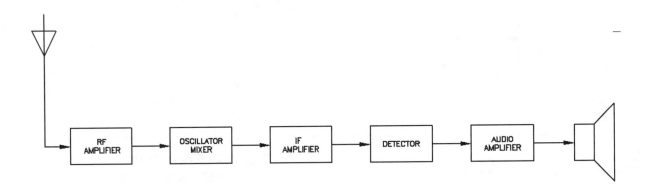

EX-ELECT-2

INSTRUCTIONS: SCHEMATIC

1. Start a **NEW** file and select **My Decimal Setup**

2. Select the **A Size** layout tab

4. Set: Snap = .100   Grid = .100.

5. Draw the schematic (shown on the next page) in Model Space

    The size is not critical but maintain good proportions and drawing balance.

6. Draw the solder points using **Donuts**. .00 I.D. and .100 O. D.

    Use layer **Circuit**

7. Add the designators, use layer **Designator**

    Text Ht = .125

8. Add the Parts List (in paperspace)

    Rows = .25   Text Ht = .125   (Try tables)

9. **Save** the drawing as: **EX-ELECT-2**

10. Plot using Plot page Setup: **Plot Setup A**

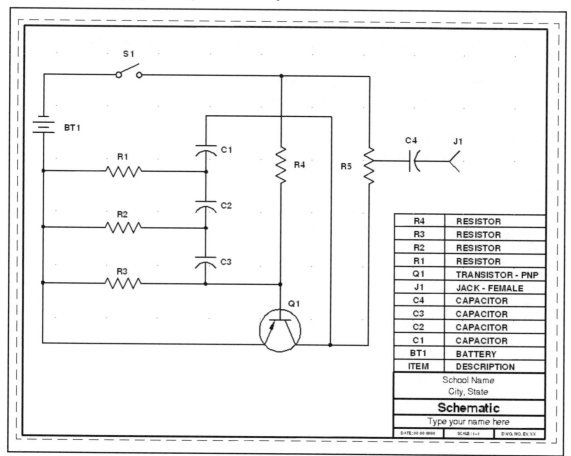

ITEM	DESCRIPTION
R4	RESISTOR
R3	RESISTOR
R2	RESISTOR
R1	RESISTOR
Q1	TRANSISTOR - PNP
J1	JACK - FEMALE
C4	CAPACITOR
C3	CAPACITOR
C2	CAPACITOR
C1	CAPACITOR
BT1	BATTERY

School Name
City, State

Schematic

Type your name here

DATE: 00 00 0000   SCALE : 1 - 1   DWG. NO. EX XX

ELECT-11

EX-ELECT-3

INSTRUCTIONS: GRID LAYOUT

Refer to drawing on the next page.

This drawing is only a template. It will be used as a template for the following 4 drawings. If you follow the instructions and draw this template correctly, the next 4 drawings will be very easy.

1. Start a **NEW** file and select **My Decimal Setup**

2. Select the **A Size** tab and unlock the viewport.

3. Adjust the Viewport scale to 2 : 1

4. Set: SNAP = .100   GRID = .100

5. Draw the board outline <u>shown on the next page </u>(on layer PCB) Full Scale (2.50 x 2.00).
    (Note: it will appear larger because you set the VP scale to 2 :1)
    Printed Circuit Boards are generally drawn 2, 10 or even 100 times larger than their actual size.

6. Draw the **Circuit** on layer **Circuit.**

7. **INSERT** symbols 19, 28, and   31 on Layer **Components** (Refer to Library)

8. Draw the designators on Layer **DESIGNATOR**.
    Text Ht = .125 (use an annotative text style and set)

9.   Draw dimensions on Layer **DIMENSION**.
    (Use an annotative dimension style)

10.   **Edit** the title block

11.   Save as **EX-ELECT-3 (Do not plot)**

EX-ELECT-3

GRID LAYOUT

Refer to instructions on the previous page.

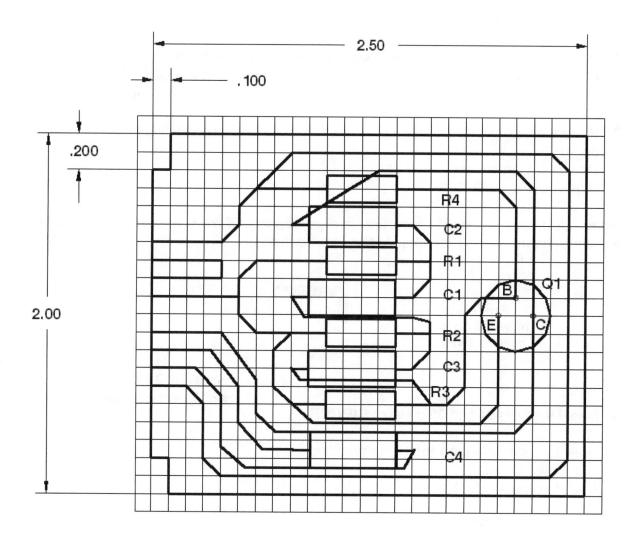

EX-ELECT-4

INSTRUCTIONS: ARTWORK

Refer to the drawing on the next page.

The following is an example of a how you might illustrate the ARTWORK for the circuit on the PCB.

1. Open **EX-ELECT-3**

2. **FREEZE** Layers **DESIGNATOR**, **DIMENSION** and **COMPONENTS**.
    (Remember, if you were not careful when you created EX-ELECT-3 the wrong objects may disappear. You may have to move objects to the correct layer)

3. Draw the heavy lines outside the corners of the PCB. (This defines the edges of the board)
    Use layer Cornermark.
    Use Polyline, width .100

Note: Consider using Offset to create a guideline, .05 from the edge of the board, for the polyline. Then just snap to the intersections.

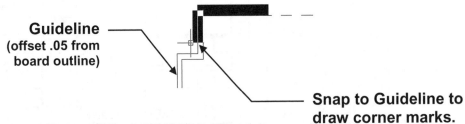

Guideline ──
(offset .05 from
board outline)

Snap to Guideline to draw corner marks.

4. **Insert** symbols **34** and **35** on layer **CIRCUIT**.

5. Draw the circuit lines **on top** of the previous lines.
    Use Layer Circuit
    Use Polyline, width .025

6. Turn **OFF** Layer **PCB**.

7. Save as **EX-ELECT-4**

8. **Plot** using Page Setup: **Plot Setup A.**

EX-ELECT-4

ARTWORK
Refer to instructions on the previous page.

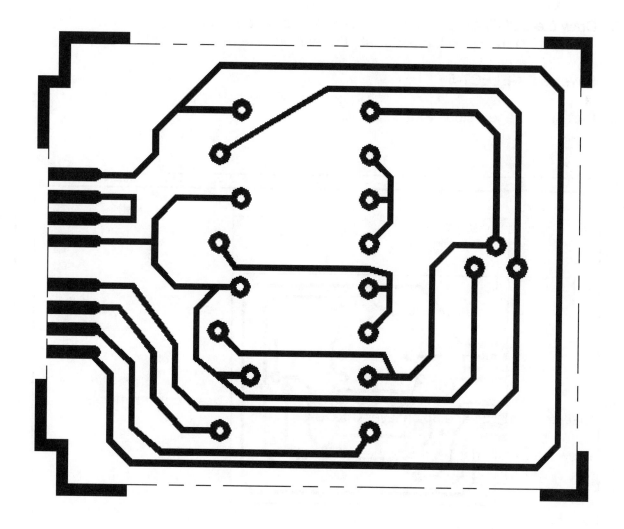

EX-ELECT-5

INSTRUCTIONS: PCB DETAIL

This exercise illustrates how you can create a drawing merely by freezing and thawing existing layers.

1. Open **EX-ELECT-4**

2. **Freeze** layer <u>Cornermark</u> and **thaw** layers <u>PCB</u> and <u>Dimension</u>

3. Draw the parts list in paperspace. Use Layers Border and Text.
   (Do not dimension the parts list. Dimensions are reference only.)

4. Save as **EX-ELECT-5**

5. **Plot** using Page Setup: **Plot Setup A.**

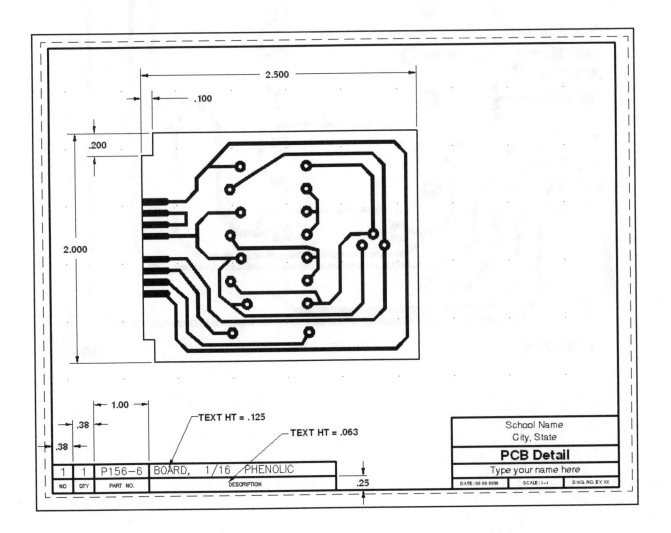

NO	QTY	PART NO.	DESCRIPTION	
1	1	P156−6	BOARD, 1/16 PHENOLIC	

TEXT HT = .125
TEXT HT = .063

School Name
City, State

PCB Detail

Type your name here

| DATE : 00-00-0000 | SCALE : 1−1 | DWG. NO. EX XX |

EX-ELECT-6

INSTRUCTIONS: COMPONENT ARTWORK

Refer to the drawing on the next page.

The following is an example of a how you might illustrate the **ARTWORK** for the **COMPONENTS**.

1. Open **EX-ELECT-3**

2. **Freeze** Layer **Dimension**.
   (Remember, if you were not careful when you created EX-ELECT-3 the wrong objects may disappear. You may have to move objects to the correct layer)

3. **MIRROR** the entire board and its components.

   WHY? A printed circuit board has the circuit on one side and the components on the other. We need to show the opposite side of the board to place the components..

   First, lets review the MIRRTEXT command you learned in the Beg. workbook.

   a. At the command line type: **MIRRTEXT <enter>**
   b. Type: **0 <enter>**   (0 means OFF,   1 means ON)

   The **MIRRTEXT** command controls whether the TEXT will mirror or not. It will change positions with the object but you can control the "Right Reading". In this case, we do not want the text to be shown reversed, so we set the MIRRTEXT command to "0" OFF.

4. ALIGN the designators.

5. Save as: **EX-ELECT-6**

6. **Plot** using Page Setup: **Plot Setup A.**

EX-ELECT-6

COMPONENT ARTWORK

Refer to instructions on the previous page.

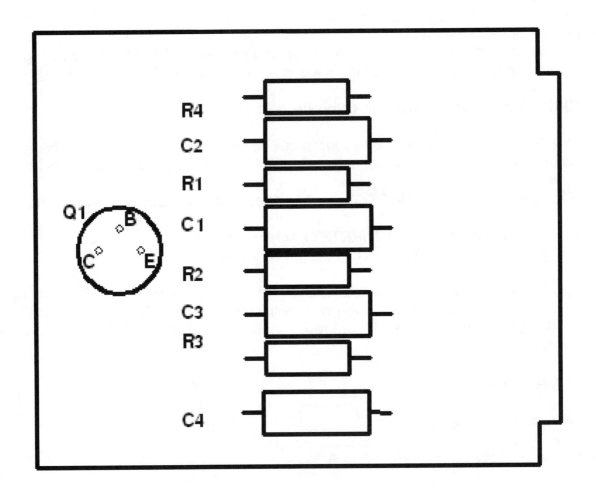

EX-ELECT-7

INSTRUCTIONS: PCB Assembly

1. Open **EX-ELECT-6**

2. Make the modifications to the drawing

3. Add the Parts List in paper space as shown.

4. Save as: **EX-ELECT-6**

5. **Plot** using Page Setup: **Plot Setup A.**

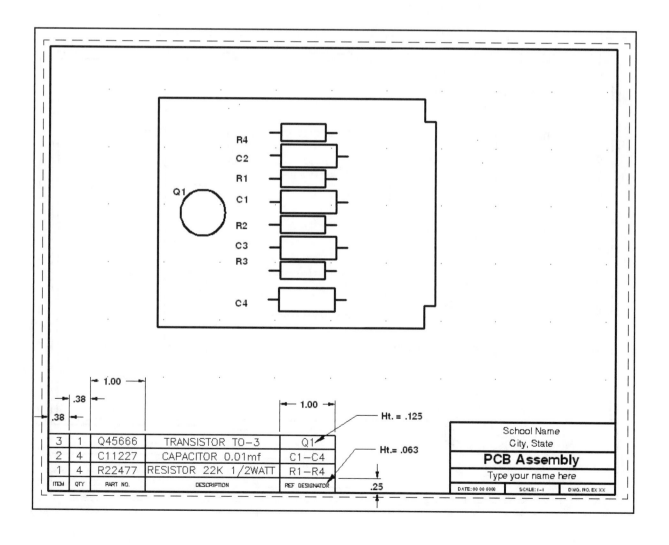

ITEM	QTY	PART NO.	DESCRIPTION	REF DESIGNATOR
3	1	Q45666	TRANSISTOR TO-3	Q1
2	4	C11227	CAPACITOR 0.01mf	C1-C4
1	4	R22477	RESISTOR 22K 1/2WATT	R1-R4

School Name
City, State

PCB Assembly

Type your name here

| DATE: 00 00 0000 | SCALE: 1-1 | DWG. NO. EX XX |

EX-ELECT-8

INSTRUCTIONS: CHASSIS

1. Start a **NEW** file and select **My Decimal Setup**

2. Adjust the Viewport Scale to 1 : 1.

3. Draw the **Flat Pattern** and dimension using Ordinate dimensioning as shown.

4. Cut an elliptical Viewport and adjust the scale to 1 : 4.

5. Draw the **Isometric** view to illustrate the **After Forming** appearance.

6. Draw the **Hole Chart** in paperspace. Size is your choice. (Try Tables)

7. Save as: **EX-ELECT-8**

8. **Plot** using Page Setup: **Plot Setup A**

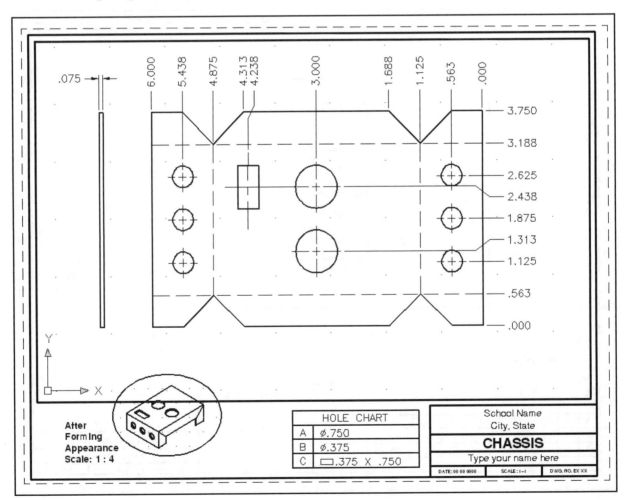

MECHANICAL

MECHANICAL SYMBOL LIBRARY

When you are using a CAD system, you should make an effort to ONLY draw an object once. If you have to duplicate the object, use a command such as: Copy, Array, Mirror or Block. Remember, this will make drawing with CAD more efficient.

In the following exercise, you will create two mechanical symbols that you use frequently. You will create them once and then merely drag and drop them, from the DesignCenter, when needed. This will save you many hours in the future.

Save this library file as **Library** so it will be easy to find when using the DesignCenter or create a Library Palette.

1. Start a New file and select **My Decimal Setup**

2. Select the **model** tab.

3. Draw each of the Symbol objects, shown on the next page, actual size. Do not scale.

4. Create an individual Block for each one using the **BLOCK** command.

5. Save this drawing as: **Library**

6. Plot a drawing of your library symbols for reference. The format is your choice.

7. Plot using Page Set up: **Plot Setup A.**

SYMBOL	INSTRUCTIONS

SYMBOL

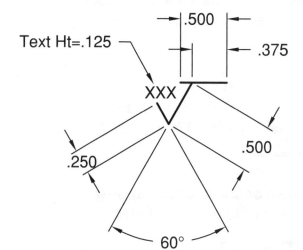

Text Ht=.125

.500

.375

XXX

.250

.500

60°

FINISH MARK

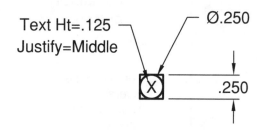

Text Ht=.125
Justify=Middle

Ø.250

.250

X

NOTE IDENTIFIER

INSTRUCTIONS

Layer = Symbol
Assign Attributes to the Text

Create a New Border

Before you can work on the exercises you must first create a new Page Setup and Border for a larger sheet of paper. Follow the following steps.

PAGE SETUP

A. Open **My Decimal Setup**.

B. Select a **LAYOUT2** tab.

B
If you do not have these tabs refer to page 2-17

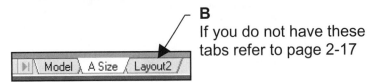

Note: If the "Page Setup Manager" dialog box shown below does not appear automatically, right click on the Layout tab, then select Page Setup Manager.

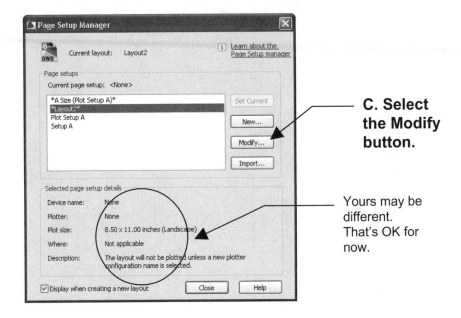

C. Select the Modify button.

Yours may be different. That's OK for now.

The "Page Setup" dialog should appear.

D. Select DWF6ePlot.pc3

E. Select ARCH (24.00 x 18.00 Inches)

F. Select Layout

G. Plot offset 0

H. Select the OK button.

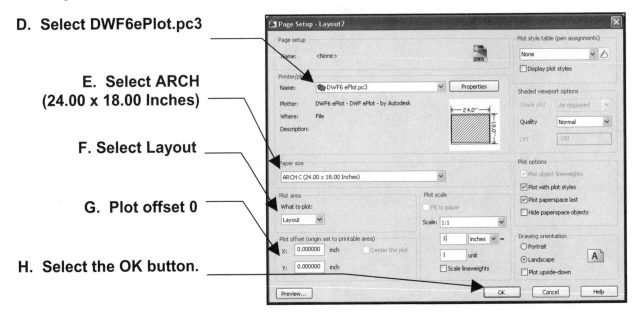

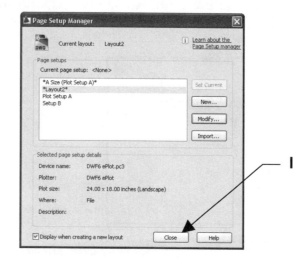

I. Select the **Close** button.

You should now have a sheet of paper displayed on the screen.
This sheet is the size you specified in the "Page Setup".
This sheet is in front of the drawing that is in Model Space.
The dashed line represents the printing limits for the device that you selected.

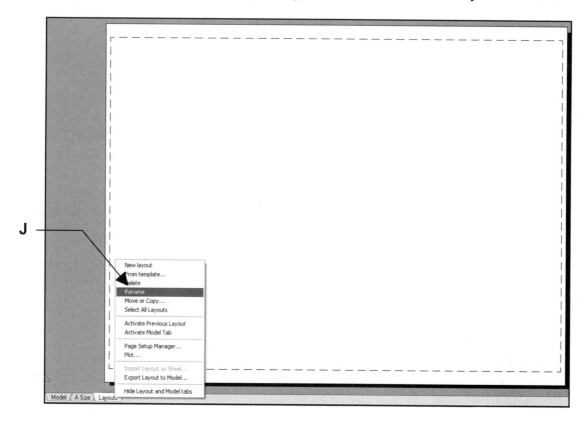

J. Right click on the Layout tab and select **Rename**.

K. Type the new name **C Size**

L. Now we need to draw a new larger RECTANGLE, shown below, to fit on the larger sheet of paper shown on the screen. (Use Layer = Border)

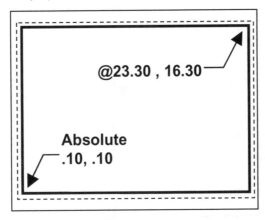

@23.30 , 16.30

Absolute
.10, .10

Now, you do not want to draw the A-Size title block all over again. The next step is to create a block from the A-Size title block and then insert it into this layout. This is how thinking ahead saves you time.

M. Select the **"A Size"** Layout tab.

    1. You need to be in "Paper Space".
    2. Select the **Make Block** command.

        a. Make a BLOCK of the "Title Block" ONLY. Do not include the Border Rectangle. <u>Notice where the "Basepoint" is selected below</u>.

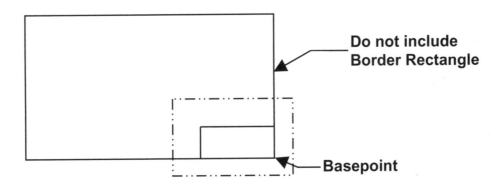

Do not include Border Rectangle

Basepoint

N. Select the **C Size** Layout tab.
    1. **Very important:** Change to Layer = Border.
    2. Select **INSERT BLOCK** command
    3. Select the "Title block" block from the list of blocks then select OK.
    4. Insertion Point should be the lower right corner of the Border Rectangle on the screen.

Do not scale the inserted Block. It may look smaller because the border rectangle is larger than the previous border, but the title block is the same size it was in A Size layout.

O. Create a Viewport approximately as shown below.

P. Save as: **My Decimal Setup.dwt**

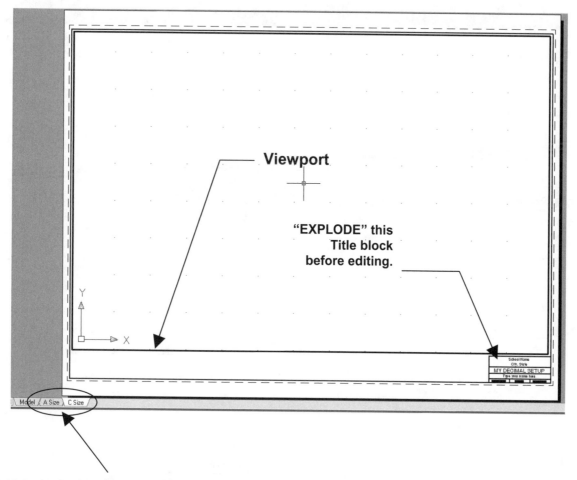

Now you have 2 master layout borders. One to be used when plotting on a sheet size 8-1/2 x 11 and one to be used when plotting on a sheet size 24 x 18. But you are not quite done yet. Next you need to set the plotting instructions. Continue on to next page.....

Q. Create new **Plot Page Setup**
   1. Select **Plot**
   2. Change the settings to match settings shown below:

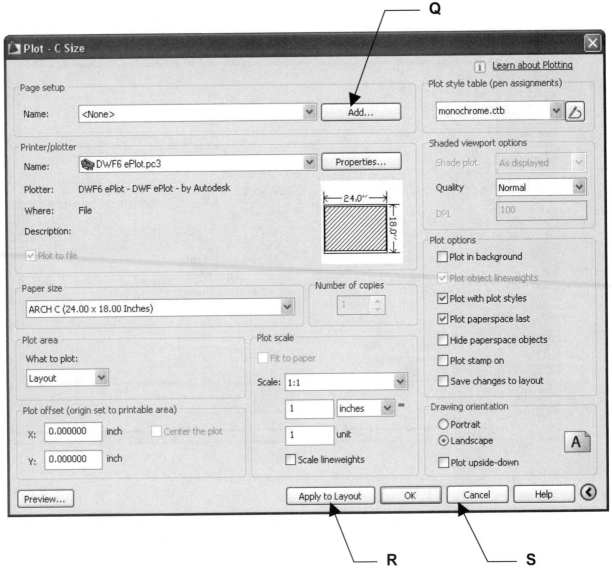

R. **ADD** a page setup name: **Plot Setup C**

S. **Apply to Layout**

T. Select **Cancel** (This closes the dialog box but your settings are still saved)

U. Save again: **My Decimal Setup.dwt**

Now you are ready to do the Exercises.

How to make a space in a dimension line.

When inserting geometric dimensioning symbols in a dimension you need to make a space in the dimension line.

Example:

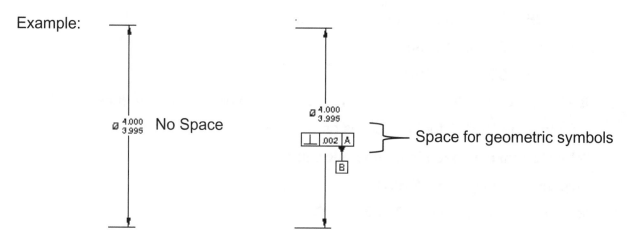

1. Place a linear dimension like you normally would.

2. Left click on the dimension and then right click and select **Properties**

3. Scroll down to: **Text** category and **Text Override**

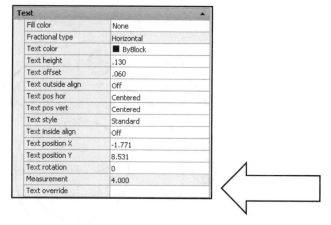

4. Type <> and the **\P** for every space you would like to enter. <> represents the **Associative** dimension. *(Note: the P must be uppercase)*

Example: For the dimension above it would look like this:

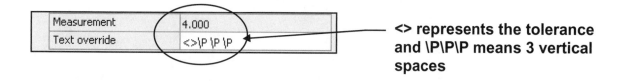

<> represents the tolerance and \P\P\P means 3 vertical spaces

EX-MECH-1

INSTRUCTIONS:

1. Start a **NEW** file and select **My Decimal Setup**

2. Select the **C Size** layout tab

4. Draw the **ECCENTRIC HUB** shown below.

5. Use Layers **Object line** and **Dimension** with geometric dimensions. (Lesson 12)

6. Refer to page Mech-9 for instructions on:

    "How to make a space in a dimension line".

7. **Save** the drawing as: **EX-MECH-1**

8. Plot using Plot page Setup: **Plot Setup C**

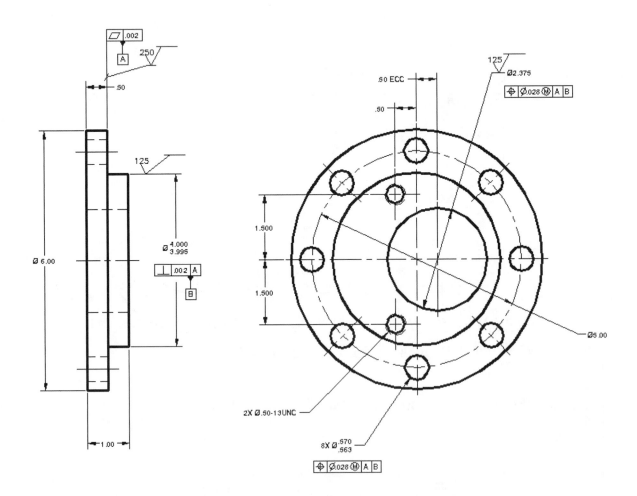

EX-MECH-2

INSTRUCTIONS:

1. Start a **NEW** file and select **My Decimal Setup**
2. Select the **C Size** layout tab
4. Draw the **SPLINE HUB** shown below.
5. Use Layers **Object line, Hatch, Hidden line, Center line** and **Dimension**
6. Refer to page Mech-9 for instructions on:

   "How to make a space in a dimension line".
7. **Save** the drawing as: **EX-MECH-2**
8. Plot using Plot page Setup: **Plot Setup C**

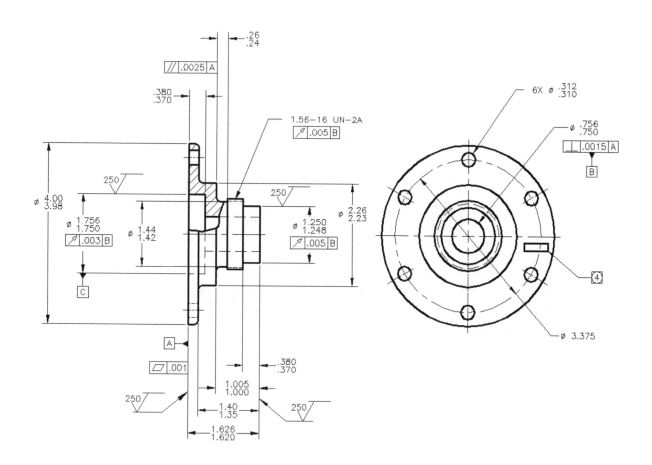

EX-MECH-3

INSTRUCTIONS:

1. Start a **NEW** file and select **My Decimal Setup**

2. Select the **C Size** layout tab

4. Draw the **Guide Ring** shown below.

5. Use Layers **Object line, Hatch, Hidden line, Center line** and **Dimension**

6. Refer to page Mech-9 for instructions on:

   "How to make a space in a dimension line".

7. **Save** the drawing as: **EX-MECH-3**

8. Plot using Plot page Setup: **Plot Setup C**

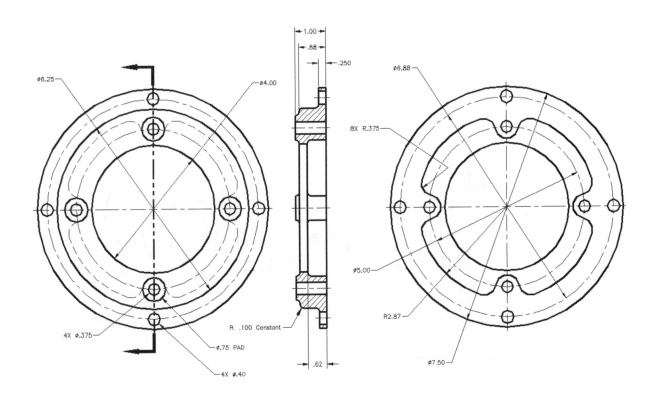

APPENDIX A
Add a Printer / Plotter

The following are step-by-step instructions on how to configure AutoCAD for your printer or plotter. These instructions assume you are a single system user. If you are networked or need more detailed information, please refer to your AutoCAD "Info. Center" (Help Index).

Note: You can configure AutoCAD for multiple printers. I suggest that you configure the plotter shown below to match the exercises in this workbook.

A. Select **Output tab / Plot panel / Plotter Manager**
B. Select **"Add-a-Plotter"** Wizard

C. Select the **"Next"** button.

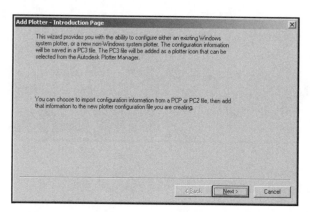

D. Select "**My Computer**" then **Next**.

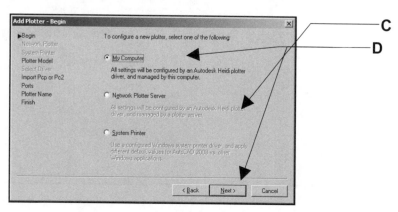

E. Select the **Manufacturer** and the specific **Model** desired then N**ext**.

(If you have a disk with the specific driver information, put the disk in the disk drive and select "Have disk" button then follow instructions.)

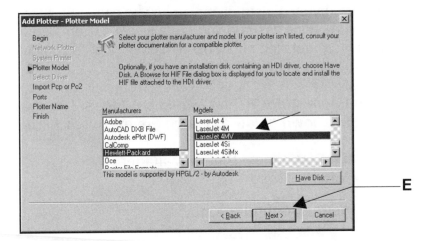

F. Select the **"Next"** box.

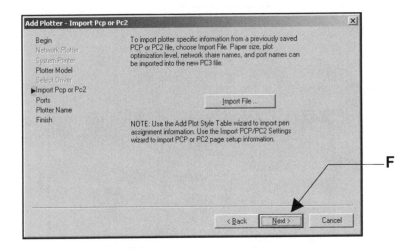

G1. Select **"Plot to a port"**.
G2. Then select **"Next"**.

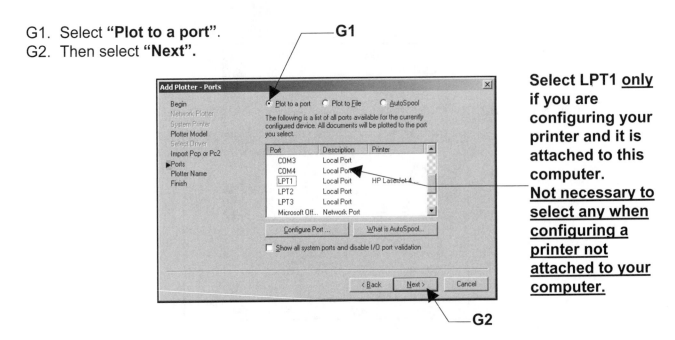

Select LPT1 <u>only</u> if you are configuring your printer and it is attached to this computer. <u>Not necessary to select any when configuring a printer not attached to your computer.</u>

H. The Printer name that you previously selected should appear.
Then select **"Next"**

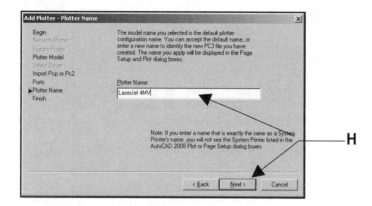

I. Select the **"Edit Plotter Configuration…"** box.

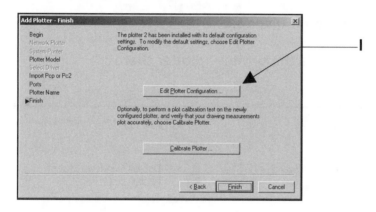

J. Select:
1. Device and Document Settings tab.
2. Media: Source and Size
3. Size: select paper size
4. OK box.

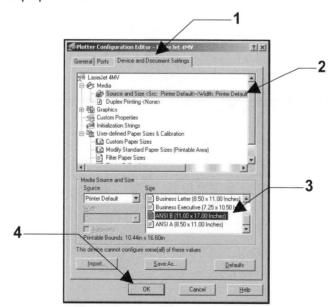

K. Select **"Finish"**.

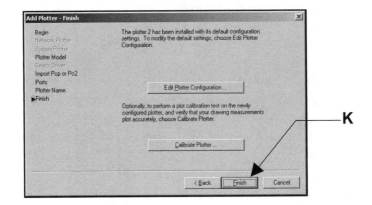

K

L. Now select the **File / Plotter Manager.**

Is the printer / plotter there?

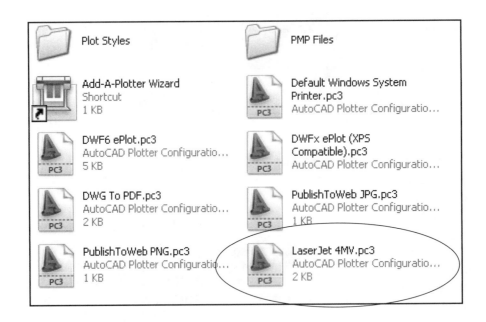

INDEX